NATIONAL DEFENSE RESEARCH INSTITUTE

T0210387

Venture Capital and Strategic Investment for Developing Government Mission Capabilities

Tim Webb, Christopher Guo, Jennifer Lamping Lewis, Daniel Egel

Prepared for the Office of the Secretary of Defense

The research described in this report was prepared for the Office of the Secretary of Defense (OSD). The research was conducted within the RAND National Defense Research Institute, a federally funded research and development center sponsored by OSD, the Joint Staff, the Unified Combatant Commands, the Navy, the Marine Corps, the defense agencies, and the defense Intelligence Community under Contract W74V8H-06-C-0002.

Library of Congress Cataloging-in-Publication Data is available for this publication.

ISBN: 978-0-8330-8213-8

The RAND Corporation is a nonprofit institution that helps improve policy and decisionmaking through research and analysis. RAND's publications do not necessarily reflect the opinions of its research clients and sponsors.

Support RAND—make a tax-deductible charitable contribution at
www.rand.org/giving/contribute.html

RAND® is a registered trademark.

Cover images: andrea1971, alisonhancock (Fotolia.com)

RAND OFFICES
SANTA MONICA, CA • WASHINGTON, DC
PITTSBURGH, PA • NEW ORLEANS, LA • JACKSON, MS • BOSTON, MA
CAMBRIDGE, UK • BRUSSELS, BE
www.rand.org

Preface

The objective of this project was to compare and contrast the organization, operation, and incentive structures of U.S. government venture capital and strategic investment initiatives. The research team conducted brief case studies of three prominent examples of such initiatives, and also created an economic model to begin to systematically describe some of the incentives underlying this type of government activity. The case studies of In-Q-Tel, the Rosettex Technology and Ventures Group, and OnPoint Technologies were conducted through literature reviews and interviews with principals in the three organizations. The research did not evaluate historical operating information sufficient to judge whether each initiative met its objective.

The summary observations about government strategic investment methods have value for understanding how previous initiatives operated, and lessons from these experiences might be applied in the design and management of future such initiatives. This report should therefore be of interest to decisionmakers, policymakers, and researchers working on the relationship between incentives provided by public and private sources for technology innovation.

This research was sponsored by the Joint Improvised Explosive Device Defeat Organization (JIEDDO) and conducted within the Acquisition and Technology Policy Center of the RAND National Defense Research Institute, a federally funded research and development center sponsored by the Office of the Secretary of Defense, the Joint Staff, the Unified Combatant Commands, the Navy, the Marine Corps, the defense agencies, and the defense Intelligence Community.

For more information on the RAND Acquisition and Technology Policy Center, see http://www.rand.org/nsrd/ndri/centers/atp.html or contact the director (contact information is provided on the web page).

Contents

Figures

Tables

Summary

The Chief of the Research and Engineering Division at the Joint Improvised Explosive Device Defeat Organization (JIEDDO) asked RAND to review previous experience with U.S. government–sponsored venture capital initiatives, and the ways they developed mission-oriented capabilities. In particular, he was interested in understanding how they were organized and how they operated, and whether and how they influenced private-sector actors to adapt their innovations for government purposes. In response to this request, RAND performed a two-pronged analysis of government venture capital and strategic investment methods. One track examined the practical experience of three prominent government venture capital/strategic investment initiatives. The second track created a game-theoretic model to explain the balance of selected economic incentives that such initiatives can use to spur innovation. The microeconomic analysis of the second track helps to shed light on the differences observed in the examples of the first track, and it lays an analysis foundation that can be used to structure and manage elements of future initiatives.

In the first track, RAND identified three recent instances of U.S. "government venture capital" initiatives that provided the most significant support for discovery and development of new technology based mission capabilities: In-Q-Tel (IQT), the Rosettex Technology and Ventures Group (RTVG), and OnPoint Technologies (OPT). These are ***government*** in their being federally chartered or operated under contract with the U.S. government. They are ***venture capital*** because private investment was intended to be an important component of their business model. They are ***strategic*** in having focused on achieving the long-term aims and interests of mission-focused U.S. government agencies. Each of these initiatives provided innovative private companies with financial support and advice so they could tailor emerging commercial products and service offerings to address a mission objective of the U.S. government. This mission-oriented *strategic* investment purpose distinguishes these initiatives from private venture capital firms, and this report subsequently uses the term *government strategic investment* (GSI) rather than "*government venture capital*." The case studies of IQT, RTVG, and OPT were conducted through a combination of literature reviews and discussions with individuals directly involved.

In the second track, RAND assumed a GSI formed with a given budget, and built an economic model to explain how the government should think about one specific policy element: the portion of government funding reserved for development of prototypes relevant to government. Although numerous other incentives influence the development of innovations within GSI (e.g., market sizes, information rights), the model focuses solely on the one element described.

Qualitative analysis of cases led to the following observations:

1. **In the three GSI cases examined, mission-oriented innovation was of equal or greater importance than generating financial return.** In each instance, significant effort was devoted to creating an organizational and legal framework that would provide direct benefit to accomplishing government mission objectives. Each GSI expended substantial effort to establish and maintain a good impedance[1] match between the private company providing the solution and its U.S. government investor/customer. In each case, there was an investment management organization to facilitate the match, turn expressions of customer need into investment proposals, conduct due diligence, and manage resources for both investment and development work programs.

2. **GSI participation in venture capital investments has provided government with additional information about technology-focused market sectors and companies.** The ability to participate directly in risk capital transactions has allowed the GSI investment managers to become part of the information sharing between entrepreneurs and private venture capitalists. The degree to which this information has translated into effective adoption of new technologies varies by case, and is not specifically evaluated in this research.

3. **GSI initiatives rely on the operational flexibility afforded by Other Transaction (OT) authority** (or "OT-like" authorities in the case of IQT) as a statutory foundation for both the contractual relationship with their sponsoring government agency investor, and the contractual relationship they enter with private companies. OT authorities have allowed GSI investment managers great flexibility to combine investment with mission need-oriented prototype programs in ways that are specifically suited to the needs of individual companies—on matters ranging from accounting practices and financial reporting to payments and intellectual property rights. Specific care is taken to structure the flow of government information rights consistent with Federal Acquisition Regulations to facilitate eventual scale adoption of prototypes.

4. **GSI initiatives rely in a significant way on a government-to-private sector "interface" function** that performs one or more of the following tasks: (1) providing contract administration, (2) identifying and "translating" investor mission-oriented needs into a form suitable for use by GSI investment managers, and (3) facilitating scale adoption of private company prototype solutions by mission-oriented government customers. These interface functions are performed by government employees and serve to ensure that inherent government responsibilities dovetail appropriately with responsibilities discharged by GSI managers. GSI personnel do not have the organizational knowledge or breadth of expertise to assimilate all potential customer needs, and the interface organization typically includes employees of the government agency investor.

5. **The GSI's responsibility to government customers adds significant difficulty to the task of investment management.** The GSI must not only serve routine investment portfolio functions, such as identifying opportunities and negotiating and monitoring investments, but must also facilitate a good impedance match between government customers and the private companies in which the GSI invests, and do so in a way that does not confuse public and private responsibilities.

---

[1]   Impedance matching is the process of designing the input of a destination component to maximize power transfer from a source component. The term has specific technical meanings in electrical engineering, acoustics, optics, and mechanics, but can be applied to any situation where energy is transferred from a source to a destination.

6. **GSI need staff with private market capabilities to serve investment management functions.** It is difficult to overstate the importance of the quality and experience of the GSI investment management personnel. They serve a variety of functions, from devising investment hypotheses to monitoring and harvesting investments. Even though staff monetary compensation for some GSI is lower than in private venture capital, the credibility necessary for staff to operate as peers in private investment transactions depends on them having skills equivalent to their private sector counterparts.

Economic modeling analysis led to the following observations:

1. **It is possible to systematically assess selected incentives to which private firms will respond** by channeling incremental technology development efforts toward government-specific prototypes—as opposed to private sector–specific prototypes. This suggests a method for resource allocation that can be used both to design aspects of future GSI and to choose among incentive mechanisms, depending on the degree to which the government and the innovating firm are sensitive to the specificity of an envisioned prototype. These sensitivities are likely to vary by technology and mission application area.

2. **The desired balance of GSI financial support between equity investment and contractual support depends on likelihood of sale in government and commercial markets.** In some situations, there will be a large difference in the likelihoods of selling a particular innovation to government versus commercial customers. In other instances, the difference in these likelihoods will be small. The flexibility inherent in OT authorities allows the GSI to balance its investment/contract offers to provide incentives for private companies to tailor innovation to address government mission objectives.

3. **The GSI initiatives in the case studies illustrate a range in the balance between equity investment and contractual support**, with OPT having most heavily emphasized the former, RTVG having most heavily emphasized the latter, and IQT having pursued a mixed strategy. Although this report does not examine the comparative effectiveness of these approaches, the economic analysis presents a framework within which to consider the suitable balance for future GSI initiatives. A more complete model would also consider incentives associated with information transfer, since these transfers are an important feature of GSI.

Abbreviations

AARCC   Advanced Agricultural Research and Commercialization Corporation

AVP   Arsenal Venture Partners

CECOM   Communications Electronics Command

CIA   Central Intelligence Agency

DARPA   Defense Advanced Research Projects Agency

DeVenCI   Defense Venture Catalyst Initiative

DoD   Department of Defense

DOE   Department of Energy

FAR   Federal Acquisition Regulation

GSI   government strategic investment

IDIQ   indefinite delivery/indefinite quantity

IED   improvised explosive device

IQT   In-Q-Tel

IT   information technology

JIEDDO   Joint Improvised Explosive Device Defeat Organization

NASA   National Aeronautics and Space Administration

NGA   National Geospatial-Intelligence Agency

NIMA   National Imagery Management Agency

NTA   National Technology Alliance

OPT   OnPoint Technologies

OT   Other Transaction

P.L.   Public Law

QIC	In-Q-Tel Interface Center
R&D	research and development
RPC	Red Planet Capital
RTVG	Rosettex Technology and Ventures Group
SBIR	Small Business Innovative Research
SIB	Social Impact Bond
VCIC	Venture Capital Investment Corporation

CHAPTER ONE

Introduction

Problem Statement

Over the past decade, the U.S. military has worked unremittingly to improve capabilities to address urgent operational needs in Iraq and Afghanistan. These include capabilities for countering enemy combatant use of improvised explosive devices (IEDs), coordinated by the Joint Improvised Explosive Device Defeat Organization (JIEDDO). A wide range of military capability improvement efforts, including those to thwart IED use, have benefited from development and procurement methods that accommodate urgent operational needs. But the decline in "overseas contingency operations" funding for military operations in South Asia has introduced planning and budgeting uncertainty into many military system acquisition initiatives, including that for countering IEDs. The adjustments in acquisition practice to this new operational tempo coincide with some threats—like that of IEDs—that are becoming persistent and commonplace worldwide. These changes in the threat environment suggest a fresh examination of the adequacy and suitability of acquisition methods for the coming decade. This study focuses on describing the government strategic investment (GSI) method as one element of this re-examination. GSI refers to situations in which government or its agent participates as a minority owner of a private firm to accomplish a government mission based on the firm's products and services. The investment is strategic in that it has significant and long-lasting positive consequences for the performance of the government mission.

The new-found persistence and globalization of the IED threat was the immediate impetus that led the Chief of the Research and Engineering Division at JIEDDO to ask a general question: What methods not currently being used might help to expand the base of technology available for developing future military capabilities? In particular, he was interested in reviewing government venture capital initiatives to understand their organization and operation, as well as to understand how they created incentives for private-sector actors to adapt their innovations for government mission applications. A number of such initiatives exist that facilitate contact between mission-oriented U.S. government agencies and innovative private companies. An important motivation for these initiatives has been the interest in reducing procedural and administrative disincentives that small private companies face while contemplating the Federal Acquisition Regulation (FAR).

This report examines the organization, operation, and incentive structure of GSI initiatives. The analysis is not specific to counter-IED missions. It does not compare the efficacy of these initiatives with one another or with other methods used by government for spurring technology innovation.

Method

The research comprised two mutually reinforcing tracks: case studies of three GSI initiatives, and an economic model that selectively helped identify and examine elements of design for GSI. The case studies relied on a combination of literature reviews, interviews with participants, and the knowledge and experience of the RAND team. The economic model is a simplified two-stage game theoretic model that addresses the relationship between GSI and a single innovating firm. In the first stage, given a fixed budget, the GSI enterprise chooses whether to allocate resources to general investment in an innovating firm or to contracts for the development of government-specific prototypes. In the second stage, the firm responds by deciding what proportion of innovative products or services will be developed into prototypes for the government and what proportion will be developed for the private sector.

Both research tracks yielded insights into the archetypal characteristics of GSI initiatives and help to explain how their structure affords government the opportunity to choose a balance among resource allocation methods to influence private firm innovation behavior.

Organization of The Report

Chapter Two describes organizational models by which investors commit resources to create technology-based innovations. It introduces the concept of a GSI initiative, and places GSI in the context of a broader ecosystem of approaches and organizations that support the development of technology.

Chapter Three presents a case study analysis in which three past and present GSI initiatives are examined in detail: OnPoint Technologies (OPT), In-Q-Tel (IQT), and the Rosettex Technology and Ventures Group (RTVG).

Chapter Four extracts key features from the case studies concerning the incentives associated with GSI and the relationship of the GSI to a representative portfolio company. It then describes a game-theoretic economic model that was developed to explore the implications of GSI resource allocation.

Chapter Five summarizes observations derived from the case studies of past and present GSI initiatives.

Appendix A contains the details of the game-theoretic economic model discussed in Chapter Four.

Appendix B contains excerpts from legislation pertaining to Other Transaction (OT) authorities.

CHAPTER TWO

Strategic Investment for Innovation Support

Elements of the Innovation Ecosystem

Although this analysis will ultimately focus on strategic investment initiatives, it is important to place these in the broader context of approaches to supporting innovation. To an important degree, GSI borrows from other well-known and widely adopted innovation-support methods in use by both public and private actors. In particular, GSI makes use of equity investment in selected firms (a feature of private capital), and specific-performance government contracts (a feature in broad use for government acquisition). Table 2.1 provides a summary of the types of innovation supporters. These include both public and private institutions, and both funders and performers. The table focuses on direct-funding forms of support, and thus excludes non-financial open-source activities that have been an important source of software innovation and related economic growth in the last decade.[1] Each category of innovation supporter shows example enterprises, a brief statement of principal focus, indication of whether enterprises in the category are funders or performers, and whether those enterprises are primarily privately sponsored or government sponsored. Some categories contain enterprises that are both funders and performers (i.e., corporate research and development [R&D] laboratories). Some contain enterprises that are both government and private (e.g., prize sponsors and social impact investors).

Prize Sponsors (Funder)

The awarding of cash prizes as inducement for accomplishing specific technological purposes has a long history. Perhaps the earliest example of such a prize was the Longitude Prize, offered by the British government in the 18th century to spur development of a practical method for determining longitude for ships at sea.[2] A more recent example was the 2009 Defense Advanced Research Projects Agency (DARPA) Network Challenge (better known as the "Red Balloon" challenge), in which competing teams attempted to locate ten large red weather balloons anchored at ten different locations across the United States, as a way of exploring the potential of social networking and crowdsourcing techniques.[3]

---

[1] UNU-MERIT, *Study on the Economic Impact of Open Source Software on Innovation and the Competitiveness of the Information and Communication Technologies (ICT) Sector in the EU*, Final Report, November 20, 2006.

[2] D. Sobel, *Longitude: The True Story of a Lone Genius Who Solved the Greatest Scientific Problem of His Time*, New York: Walker and Company, 1995.

[3] Defense Advanced Research Projects Agency, *DARPA Network Challenge*, web page, undated.

Table 2.1
Summary of the Innovation Ecosystem

Innovation supporter	Example Enterprises	Principal Focus	Funder	Performer	Government	Private
Prize sponsor	Heinlein Prize Trust, Xprize Foundation, DARPA (Red Balloon)	• awards cash prizes for accomplishing specific objectives	X		X	X
Corporate R&D laboratory	IBM TJ Watson, GE CR&D, Microsoft Research, Xerox PARC, Google.org	• conducts R&D strategic to parent, with own resources	X	X		X
Non-profit contract research center	Battelle Ventures, Noblis, Scripps Research, SRI International, Howard Hughes MI	• develops commercially relevant technology and provides strategic technology advising		X		X
University R&D center	CMU Cylab, Stanford Metaphysics Lab, Berkeley Nanosciences, Wharton Customer Analytics	• conducts research with public and private sponsorship	X	X	X	X
Foundations	BMGF, W.K. Kellog, MacArthur, Lilly	• give grants, scholarships, matching gifts, and other forms of support	X			X
Public-private facilitator	DeVenCI, USAA Innovation Center	• facilitates connections between commercial providers and government customers			X	X
U.S. government R&D grantor	DARPA, NIH, NSF, SBA (SBIR)	• organizes and funds high-risk, high-reward, mission-specific research using grants and contracts	X		X	
U.S. government R&D laboratory	NRL, ARL, AFRL, LANL, LBL, ORNL	• performs government mission-specific RDT&E		X	X	

Table 2.1—Continued

Innovation supporter	Example Enterprises	Principal Focus	Funder	Performer	Government	Private
Financial investor	Accel Partners, Sequioa Ventures, Tech Coast Angels, KKR	• invests for financial return • private equity and venture capital	X			X
Social Impact investor	Shell Foundation, BMGF, MassVentures, Chesapeake Innovation, DOE Loan Programs, SBA SBIC	• invests for social good • private foundations • public corporations • business incubators • small business programs	X		X	X
Strategic investor	Intel Capital, McDonald's Ventures, In-Q-Tel, OnPoint, Rosettex, AARCC, Redplanet	• invests to create mission value for financial parent	X		X	X

SOURCE: RAND analysis.

Corporate Research and Development Laboratories (Funder and Performer)

U.S.-based R&D is part of a large and growing commitment to R&D by corporations world-wide (although there has been a decline over time in the degree to which corporate R&D is concentrated in the United States).[4] Robust examples are found in industries that include automotive, chemicals and materials, telecommunications, pharmaceuticals, and information technology. Companies like Microsoft, Novartis, Toyota, and Cisco routinely devote billions of dollars annually to the discovery and development of techniques to improve their products and services.

Non-Profit Contract Research and Development Centers (Performer)

There are many non-profit organizations that either provide R&D services to third parties (e.g., Noblis and Mitre Corporation) or that conduct research on their own account for the benefit of society (e.g., Howard Hughes Medical Institute). These organizations also often provide technology advising services to both private and public clients. Some of these organizations have played a part in building U.S. government strategic investing enterprises, such as when SRI International teamed with Sarnoff Corporation to form the Rosettex Technology and Ventures Group—the subject of one of the strategic investing case studies we examine below.

University Research and Development Performing (Mainly Performer)

Higher-education expenditures on R&D in the United States exceeded $60 billion in 2011, with over half of that total coming from federal government sources. This spending takes place in over 900 universities and includes not only R&D in the sciences and engineering, but in business and management, communications, education, and other disciplines.[5] This spending often leads to the creation of innovative small companies who subsequently receive funding from financially motivated investors, and whose innovations are directly applicable to the mission needs of U.S. government agencies and departments.

Foundations (Funder)

There are hundreds of non-profit grantmaking foundations in the United States, whose total giving in 2011 was $49 billion[6]; of this total, at least several billion dollars were awarded to individuals and institutions for R&D.[7] For example, the Bill and Melinda Gates Foundation provides financial resources for R&D into more effective treatments, diagnostics, and control measures for malaria. The MacArthur Foundation supports a research network examining the multiple effects of modern neuroscience on criminal law. The Packard Foundation has a broadly based program to support innovative approaches to managing climate change and improving the environmental performance of agriculture and biofuels production. The

---

[4]   "2012 Global R&D Funding Forecast," *R&D Magazine*, December 2011, p. 12. Eighteen U.S. corporations are among the top 50 firms according to R&D spending, and 2012 estimates of U.S. R&D performed by industry were $273 billion.

[5]   R. Britt, *Universities Report Highest-Ever R&D Spending of $65 Billion in FY 2011*, National Science Foundation, InfoBrief NSF-13-305, November 2012.

[6]   The Foundation Center, *Change in Foundation Giving Adjusted for Inflation, 1975 to 2011*, FC Stats: The Foundation Center's Statistical Information Service, 2013.

[7]   National Science Foundation, *National Patterns of R&D Resources: 2010–11 Data Update, Detailed Statistical Tables*, NSF 13-318, April 2013.

National Science Board estimated that foundations provide about 10 percent of the total of basic research support in the United States in 2008.[8]

Public-Private Facilitator

A small number of organizations, such as the Defense Venture Catalyst Initiative (DeVenCI) and the U.S. Air Force Center for Innovation, support R&D by coordinating the actions of third parties, rather than by directly funding or performing the work. DeVenCI facilitates regular interactions among the members of the private venture capital community, small innovative companies, and Department of Defense (DoD) mission managers, facilitating "transfer of knowledge and improved understanding between DoD participants with specific capability needs and small innovative companies."[9] DeVenCI has focused on facilitating the purchase of field-ready products and services by DoD, rather than on the discovery and development of new capabilities.[10]

U.S. Government Research and Development Grantor (Funder)

There are a number of organizations that provide financial grants to support pursuit of scientific and technological innovation. Some of these are mission focused (e.g., DARPA, the Intelligence Advanced Research Projects Activity, the Advanced Research Projects Agency–Energy). Others support fundamental research across a broad range of disciplines (e.g., the National Science Foundation). Some of these organizations maintain important information exchanges with financial investment and other growth capital entities (such as DARPA's Cyber Fast Track initiative[11]) although they do not make direct investments in companies. These organizations typically fund high-risk projects with potentially high rewards. The Small Business Innovative Research programs at various federal departments and agencies also fit here.

U.S. Government Research and Development Laboratories (Funder and Performer)

A system of centralized national laboratories that conduct research and technology development grew out of World War II, and these continue to be a significant component of the U.S. innovation ecosystem. This national laboratory infrastructure is quite diverse. A significant portion is managed by the Department of Energy (DOE), and includes such organizations as Sandia, Los Alamos, Lawrence Livermore, Pacific Northwest, and Brookhaven national laboratories. There are also dozens of Army, Navy, and Air Force laboratories focused on a wide array of R&D activities. There are also more than 30 federally funded research and development centers sponsored by the U.S. government that conduct basic and applied R&D.

---

[8]   National Science Board, *Research and Development: Essential Foundation for U.S. Competitiveness in a Global Economy*, Arlington, VA, 08-03, January 2008.

[9]   Defense Venture Catalyst Initiative, *Welcome to Defense Venture Catalyst Initiative*, undated.

[10]   A. McBride, *Pentagon Turns to Silicon Valley for Leads*, Reuters US Online Report, Technology News, October 14, 2011; and conversation with a venture capital manager who has been associated with DeVenCI since 2006 (September 2, 2012).

[11]   K. J. Higgins, "'Mudge' Announces New DARPA Hacker Spaces Program," *DarkReading.com*, August 4, 2011.

Detailed Focus on Growth Capital Approaches

We give the greatest attention to the last three approaches in Table 2.1, because the designers of federally sponsored GSIs have focused most on adopting features from these models. There are the three models by which private investors look to profit from innovative technology-based business activity: (1) financial, (2) social impact, and (3) strategic. These models differ primarily by the investor's objectives, and how these are reflected in incentives for exchange of value between participants. Each model relies in some way on innovative activity by a private company that receives equity investment. The investment is made—ideally—in a way that aligns the company and investor objectives, which may include financial return, accomplishment of non-monetary goals, or the gathering of information that increases competitive advantage. We briefly examine each model to provide background context for a more detailed description of strategic investing methods pursued by the U.S. federal government.

In each of the three models, there are four principal classes of actor: investors, investment managers, companies, and customers. We refer to this subsequently as the *four-actor model*. A transfer of value among these actors supports the achievement of monetary and non-monetary objectives. Investor financial liquidity can be achieved through a variety of means, including the acquisition of one business by another, through dividend and interest payments, and sale of stock to investors in public or private markets. We examine each of the models in turn.

Financial Investing

In the Financial Investing model (Figure 2.1), the investor is concerned primarily with achieving the greatest possible return on invested capital in the shortest possible time. The investor, which might be an individual, a financial institution, or an operating company, commits money that the investment manager invests in a company. Investment managers may co-mingle monies from multiple investors in a common pool or fund, which then serves as a resource base for the manager to buy ownership interests in multiple companies. The companies in which the manager invests are often referred to as "portfolio" companies. For the sake of simplicity, Figure 2.1 shows the exchanges of value as if the investment manager only invested in one company.

The legal and procedural mechanics of the arrangements between investor and investment manager can be complex, but are generally characterized by arms-length transactions

Figure 2.1
Financial Investment Growth Capital Model

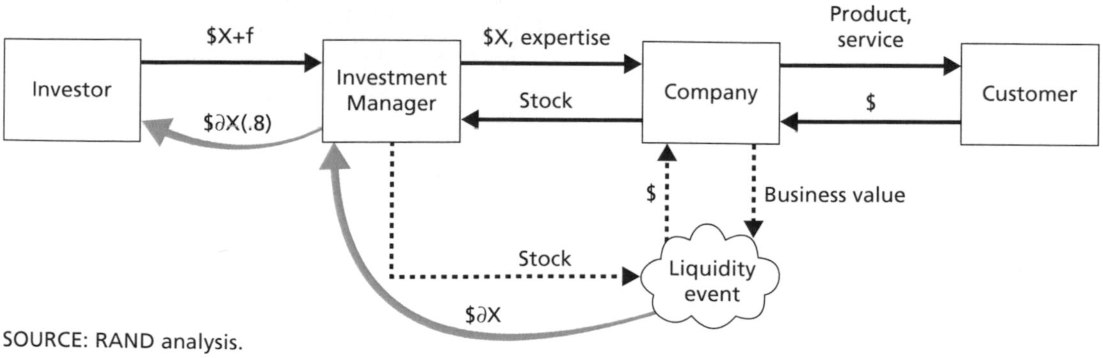

SOURCE: RAND analysis.

RAND RR176-2.1

in which investors defer to the manager's judgments for making investments, and they forgo management authority and responsibility in order to enjoy limits to legal liability. In exchange for assuming management authority and responsibility for invested funds (X), the investment manager takes a percentage fee (f) based on total capital committed to the pool or fund, and a percentage of gains or profits. In a typical venture capital arrangement, the manager retains 20 percent of investment gains resulting from a liquidity event[12] and returns 80 percent to the investor. In this example, "∂" refers to the multiple of capital that has resulted from building the value of the company; if a unit of investment initially cost the investor $1, and was worth $10 at the liquidity event, then ∂ equals 10. The "liquidity source" is the means by which the liquidity event takes place—an acquisition or public market offering that allows the investor to exit his position for cash.

Frequently, there is no relationship between the investor and the company. Although an investor will occasionally become a customer for the company's product or service, this usually involves buying the generally available commercial offering, and not tailoring that offering to a specific need.

All other things equal, a company will prefer working with an investment manager whose knowledge and experience holds the greatest potential for developing the business. The investment manager purchases stock (or debt convertible into stock) in individual companies, after negotiating valuation and other terms, such as profit participation, stock redemption rights, board membership, and voting rights. The manager will very often have prior experience in business operations, and will share his or her expertise with the leadership of the portfolio company. This is one of the attractions for an entrepreneur choosing to work with a particular investment manager—beyond the entrepreneur's interest in the investor's money, he or she also wants the company to benefit from the manager's advice and counsel.

Depending on the stage of development of the company at the time of investment, the investment proceeds may be used for purposes ranging from product prototyping and initial customer engagement to sales growth. There is a wide range of possible outcomes: A company may return capital and profits to investors without need for further rounds of investment. More often, multiple rounds of investment are necessary to develop the company to the point where a liquidity event is possible. Ideally—from the investor's point of view—there is an increase in the company's value on the occasion of each subsequent investment round. Companies may fail at any stage of this process.

Examples of organizations devoted principally to financial returns include Accel Partners, New Enterprise Associates, and Sequoia Ventures. These venture capital businesses fund companies in multiple geographies and market segments, and at stages of maturity ranging from early-stage through growth. As of 2011, there were 1,503 firms in the United States actively managing private equity funds, the large majority of which employ some form of the financial investing model.[13]

---

[12]   A liquidity event occurs when a company's stock can be sold or exchanged for cash. One example is when a company is acquired, and the acquiring company pays cash for the acquired company, or exchanges its publicly tradable stock for that of the acquired company. In this circumstance, the investor can sell or exchange their stock for cash, and thereby become "liquid." Another example is when a company's stock becomes publicly tradable through an initial public offering. By way of contrast, if the acquiring company's stock is not publicly tradable, the acquisition is not a liquidity event, because the investor cannot sell their investment position for cash.

[13]   Prequin, Ltd., *2012 Prequin Global Private Equity Report*, 2012.

There are no public or government venture capital investors who behave primarily by this model. However, there have been many efforts within the U.S. government to explore the benefit of financial investing methods for engaging with emerging companies focused on commercial markets.[14]

Social Impact Investing

In the Social Impact model (Figure 2.2), the investor views social good outcomes (e.g., job creation, community development, environmental sustainability) as having significant value. These objectives may be more important to this type of investor than financial returns. The social impact investor therefore seeks an investment manager who will achieve both social good and financial objectives. Since there are relatively few investment managers whose primary business is to achieve social good objectives, the social impact investor may choose to commit capital to a financial investment manager who is willing to make the best effort to accomplish social good objectives alongside their focus on financial returns. Alternatively (or in combination with "best efforts"), the social impact investor may host the investment management function directly within the investor organization, where it can exert a more direct influence on its operations.

The non-financial objectives of the social impact investor may result in a closer relationship between the investor and the investment manager than is common under a pure financial investing model. This is particularly true when the investor hosts the investment management function, or creates a separate investment management organization devoted to its objectives. This structure may have the concomitant result of more-routine and closer relationships among the investor, the investment manager, and the portfolio companies.

Examples of social impact investors include MassVentures and the Maryland Venture Fund, each of which was created to fill a perceived "capital gap" for start-up and expansion-stage companies. Their social impact return on investment is measured by the total capital they invest and the number of jobs created in their respective states. Figure 2.2 is a reminder that the investor values such things as jobs created by the company—although that value does

Figure 2.2
Social Impact Investing

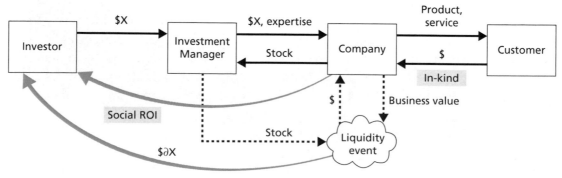

SOURCE: RAND analysis.
NOTE: ROI = return on investment.
RAND RR176-2.2

---

[14] See, for example, D. R. Graham, J. P. Bell, and A. J. Coe, *Defense Venturing Process: A Model for Engaging Venture Capitalists and Innovative Emerging Companies*, Institute for Defense Analyses, D-2847, March 2003.

not "flow" to the investor in a way that is analogous to financial returns. The financial success of the investor and manager is measured the same way as for financial investors: by return on investment. In its 30-year history, MassVentures has invested in 132 companies, 60 percent of which were located in economically targeted areas. There have been over 7,500 individuals employed in its portfolio companies, and it has generated a gross internal rate of return of 16.5 percent. Through investment gains, MassVentures has been able to generate 86 percent of the funds necessary for its continuing operations. While this financial result might not be judged adequate in a firm devoted purely to financial returns, it is more than sufficient to sustain MassVentures in its social impact focus.[15]

Organizations such as MassVentures are public or quasi-public, and were created as the result of public policy initiatives. The investment management staff is often compensated at levels lower than for counterparts in private venture capital firms, and it is less common that compensation includes the same degree of profit participation. However, the investment management staff is not always expected to contribute significant personal funds to the investment pool, as is routinely expected of managers in financial investment businesses.

An example of a privately funded social impact investor is the Shell Foundation, a UK-based charity that acts like an investor, focusing its resources on addressing global challenges linked to the use of energy. The foundation incubates new organizations (some private, some quasi-public) that provide environmentally sustainable and commercially viable approaches for such purposes as food and clothing supply. They are less interested in financial returns than in building financially self-sufficient organizations that can provide long-term assistance to people affected by economic and environmental changes stemming from the introduction of new energy technologies and supplies.[16] Social impact investors often measure success by such tangible benefits that are not derived from technology R&D.

There is also a newer type of social impact investment model, where the investors and/or service providers profit if and when desired social impacts occur. One example of this is the Greater London Authority and the £5 million "Social-Impact Bond" (SIB) that it sold to investors. Proceeds from the bond sale go to fund social service organizations whose programs serve the homeless in ways that reduce those individual's reliance on government services. Savings from reduction in demand for government services then fund payments to SIB investors, who can earn up to 6.5 percent if targets are met. Instances of this public-private partnership social investing model are spreading rapidly.[17]

These social impact investing enterprises were created to address what their sponsors perceived to be the persistent failure of private markets to address such problems as inadequate supply of capital for early-stage companies and inadequate support for organizations advancing sustainable development. Although the commercial viability of the supported companies is important to the investors, the scale and timing of financial gain is comparatively unimportant. There are several dozen active private social impact investment firms, and 10–12 other state and regional firms.[18]

---

[15] MassVentures, *History*, web page, undated.

[16] Shell Foundation, *Establishing Entities*, web page, 2008.

[17] "Social Impact Bonds: Commerce and Conscience," *The Economist*, February 23, 2013.

[18] O. Khalili, *15 Social Venture Capital Firms That You Should Know About*, Cause Capitalism, April 2, 2010.

Strategic Investing

The last model we describe is the Strategic Investing model (Figure 2.3). In this model, the investor wants direct access to one or more of goods, services, and information that will have strategic benefit for his or her organization—strategic in having significant and long-lasting positive consequences for performance of the mission or for improving business financial performance. As in social impact investing, the investor wants more than the financial returns associated with the investment itself—they want to gain knowledge from the portfolio company that will allow them to better develop their own business. The strategic investor is active in advising the company on what products or services should be offered, and they often directly influence company innovation and product development. This can be accomplished through such means as sharing technical know-how, providing access to manufacturing capability, and becoming a significant customer.

The strategic investor needs an investment management capability that intimately understands its business, so that it can effectively identify investment opportunities most likely to yield strategic value. The investor has a specific and abiding interest in maintaining contact with the companies who receive the investment, and looks to the investment manager to establish and maintain appropriate lines of communication. The investment manager is expected to provide advice and counsel to the portfolio company, but now the investor is also directly involved in supporting the company. In many cases, the investment manager is an employee of the strategic investor.

There are numerous corporate strategic investment enterprises, such as Amazon Venture Capital, T-Mobile Venture Fund, Nokia Growth Partners, Motorola Ventures, and Intel Capital.[19] These are devoted to making investments in innovative companies whose technological or business process innovations might unlock significant new business value when employed on a large scale. There have also been a small number of government-sponsored strategic investment

Figure 2.3
Strategic Investing

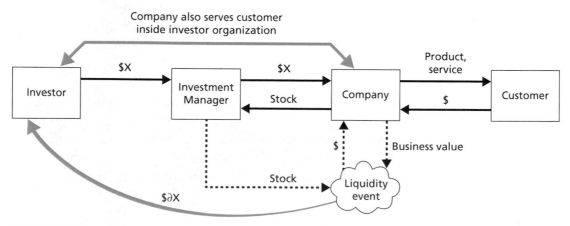

SOURCE: RAND analysis.
RAND RR176-2.3

---

[19]   T. Taulli, "The Lowdown on Strategic Investments," *Bloomberg Businessweek*, July 2, 2008. During 2012, there were over 300 investments in early-stage companies (A and B round investments) by corporate venture capital organizations. In 2012, through November, worldwide and including all stages of investment, there were 935 investments worth $16.4 billion. See Mawsonia Limited 2010, *Global Corporate Venturing Annual Review 2012*, Issue 031, December 2012.

enterprises, such as IQT, OPT, and RTVG. Their primary investment motivation is less about increasing business value than it is about improving the capabilities of government to address mission objectives.

Small innovative private companies often covet the possibility of strategic investment from such sources, not only for the prospect of becoming a customer of the investor, but more broadly for the knowledge and influence that the investor can exert to make the company successful. Although the portfolio company may simultaneously receive investment from both financial and strategic investors, the relationship with the strategic investor is often distinctive in providing access to proprietary technology, process know-how, and discerning customers. Corporate strategic investment sometimes leads to liquidity for a portfolio company—when the investor comes to believe that the business value of that company is worth acquiring in its entirety.

The strategic investor exerts influence on the portfolio company in several ways. The investment itself ($X) provides resources for the company to accelerate the achievement of current product development, marketing, and sales objectives. But the relationship also influences the evolution of the business plan over time. This may happen because the strategic investor becomes a customer for the company's product or service. The business of the strategic investor may be closely related to that of the company (e.g., Intel Capital investing in a developing semiconductor capital equipment company), in which case the investor/customer can provide the company with a much more detailed critique of its product offerings than would be available from other customers. The investment can also create expectations of frank and open communication between the company and its investor/customer, which may lead to the company adapting its technology development and product plans more quickly and effectively than would otherwise be possible. This relationship of mutual influence between investor and company is uncommon in financial and social impact investing. But it is one of the central features of strategic investing.

Improving Understanding Through Both Economic Reasoning and Case Studies

The mechanisms through which this innovation ecosystem operates have long been a focus of economics scholarship and literature. This literature particularly emphasizes the importance of patents[20] and prizes. Nalebuff and Stiglitz[21] note that "competitive" compensation for inventions is often preferable to non-competitive individual research contracts. Many economic models treat winning a prize and winning a patent as formally equivalent—the prize value being assumed to equal the expected value of profits derived from a patent.[22] But these models do not explore the circumstances under which each of the mechanisms might lead to preferred outcomes. Therefore, to supplement and complement the GSI initiative case studies of Chapter Three, RAND also created a game-theoretic model to characterize a specific element in the

---

[20] A patent is the exclusive right granted by government to an individual or organization for the use, practice, or sale of an invention for a specific period of time.

[21] B. J., Nalebuff and J. E. Stiglitz, "Prizes and Incentives: Towards a General Theory of Compensation and Competition," *The Bell Journal of Economics*, Vol. 14, No. 1, Spring 1983, pp. 21–43.

[22] S. Scotchmer, *Innovation and Incentives*, Cambridge, Mass.: MIT Press, 2004.

design of a GSI initiative that drives interaction between agents in the strategic investment process. Exercising this model helped to generate insights into the similarities and differences between the GSI cases.

These two complementary analysis tracks are presented in the next two chapters. Chapter Three presents the case studies, and Chapter Four presents the comparative economic model.

CHAPTER THREE

Case Studies of U.S. Government Strategic Investment

Identifying Candidates

There have been a number of GSI initiatives over the past 20 years, some more enduring and successful than others. Table 3.1 shows the examples that the research team identified as having made or planned to make equity investments in innovative young companies—both to enhance the mission capabilities of government and to generate financial returns.[1] Each will be described briefly before narrowing consideration to three instances in which the initiative was of a scale and duration that allowed for identification of useful lessons. There is a short but important digression in the middle of the chapter to describe the unique form of statutory authority under which these GSI operated, without which it is impossible to fully understand their workings.

Advanced Agricultural Research and Commercialization Corporation (AARCC)

The AARCC "was established in March 1992 as an independent entity within the U.S. Department of Agriculture." AARCC was created "to expedite the development and market penetration of non-food, non-feed value-added industrial products from agricultural and forestry materials and animal by-products." It provided equity investment to startup firms that were "seeking to introduce and make viable innovative enterprises to utilize agricultural products and create employment."[2] Unfortunately, a 1999 government audit of AARCC revealed that the selection and monitoring of investments had not been well managed. In one example, a portfolio company got into a protracted legal dispute over royalty rights associated with manufacturing multiple products on a common equipment base. The audit suggested that there were multiple such examples of investment agreements that had not been carefully negotiated

---

[1] This selection excludes some organizations and program initiatives that have been characterized by other scholars as constituting government "venture capital," including the Small Business Innovative Research (SBIR) programs of the various federal agencies, the VC@Sea program of the U.S. Navy, DeVenCI, and the Army's Applied Communication and Information Networking. For more on some of these organizations, see C. Brown, P. Winka, and H. Lee, *Government Venture Capital: Centralized or Decentralized Execution*, Naval Postgraduate School, NPS-AM-07-052, January 17, 2008. The selected activities are not venture capital in that they do not involve the government or its agent holding an ownership position in private companies. Venture capital investors typically receive rights to information about the entirety of private company plans, operations, and intellectual property. In the context of SBIR grants, the federal government typically has information rights more limited to operations and intellectual property relating to specific performance grants.

[2] D. Smith, *USDA Investments at Risk Due to Corporation's Mismanagement*, U.S. Department of Agriculture press release, December 2, 1999.

Table 3.1
Candidates for Case Examination

	Years	Markets Addressed	Legal Form	Relationship with Venture Capital Community	Notes
Advanced Agricultural Research and Commercialization Corporation	1992–2000	Agriculture	Independent within USDA	Limited	Terminated for lack of effective internal control structure, limited effectiveness
In-Q-Tel	1999–Present	Multiple	Non-profit	Strong, various markets	Evolving emphasis on strategic investing has supported an evolving set of investment themes
OnPoint Technologies	2003–Present	Portable power	Non-profit	Strong, sector focused	Focus on man-portable power solutions for field deployment with soldiers
Red Planet Capital	2006–2007	Space-related	Non-profit	Strong, sector focused	Did not advance far beyond inception, made only one investment
Rosettex Technology and Ventures Group	2002–2009	Imaging, geo-spatial	Private joint venture	Modest	Built partnerships for NIMA/NGA. Not renewed in 2009

SOURCE: RAND analysis; Department of Agriculture, Office of the Inspector General, *Assessment of the Alternative Agricultural Research and Commercial Corporation–Management Lacking Over High Risk Investments*, Audit Report No. 37099-1-FM, November 1999; Defense Venture Catalyst Initiative, *Frequently Asked Questions*, undated; Business Executives for National Security, *Accelerating the Acquisition and Implementation of New Technologies for Intelligence: The Report of the Independent Panel on the Central Intelligence Agency In-Q-Tel Venture*, June 2001; B. Held and I. Chang, *Using Venture Capital to Improve Army Research and Development*, Santa Monica, Calif.: RAND Corporation, IP-199, 2000; conversation with former senior executive, National Aeronautics and Space Administration, January 28, 2012; M. Reardon and D. Scott, *Rosettex NTA Project Portfolio, Final Edition*. Sarnoff Corporation, National Technology Alliance, TR-001-072709-554, September 2009; Defense Venture Catalyst Initiative, *Department of Defense Launches the Defense Venture Catalyst Initiative to Speed Discovery of Emerging Commercial Technologies*, undated.

or monitored. The only material lesson we draw from AARCC is that GSI initiatives must be skillfully managed, not only to protect the government's interests, but also to yield valuable mission outcomes. AARCC is not examined further here.

In-Q-Tel (IQT)

IQT was established in 1999 as a non-profit corporation in response to an appreciation on the part of senior officials in the Central Intelligence Agency (CIA) that there was a significant gap between understanding and implementation of information technology (IT) in the CIA, and similar practice in the private sector.[3] IQT "was designed to be an agile, flexible commercial firm that could work on its own terms with firms in Silicon Valley and throughout the world."[4] Since inception, IQT has built technology and investment practices in many subject areas, including application software and analytics, embedded systems and power, digital identity and security, and physical and biological science. Its investment themes have evolved through time to meet changing mission needs. IQT began by working exclusively for the CIA, but it has subsequently partnered with a number of other agencies inside and outside the Intelligence Community, including the Defense Intelligence Agency, National Security Agency, National Reconnaissance Office, National Geospatial-Intelligence Agency (NGA), Federal Bureau of Investigation, Department of Homeland Security, and the Transportation Safety Administration.[5]

The business model of IQT includes both equity investments in portfolio companies and funding of contracts leading to prototype demonstrations for prospective customers inside the CIA and its partner agencies.

OnPoint Technologies (OPT)

OPT "was founded by the U.S. Army in 2002 to address the Service's continued need for new power and energy solutions."[6] OPT is "specifically interested in companies that do not normally do business with the government," and "helps these companies transfer technologies to better equip soldiers and/or reduce the costs associated with such equipment."[7] OPT is a non-profit organization managed by a for-profit company named Arsenal Venture Partners (AVP); AVP also separately provides assistance to the Army for the Small Business Innovative Research (SBIR) Commercialization Pilot Program. Although OPT does not provide financial support for mission-specific developments as frequently as does IQT, they describe themselves as more risk tolerant than a commercial venture capital firm precisely because their primary goal is a mission solution for the Army.

---

[3]   See W. Molzahn, "The CIA's In-Q-Tel Model – Its Applicability," *Acquisition Review Quarterly*, Winter 2003. In this paper, the author argues for establishing a "venture catalyst firm" to serve DoD in the same way that IQT serves the CIA. Also see A. Laurent, "Raising the Ante: Venture Capitalists Are Helping Government Buy Its Way Back into the Emerging Technology Market," *Government Executive*, June 1, 2002.

[4]   Molzahn, 2003, p. 49.

[5]   A. Pratt, *Innovation for the U.S. Intelligence Community*, Summer Intern Blog, Stanford Graduate School of Business, July 14, 2011.

[6]   A. S. Mara, *Maximizing the Returns of Government Venture Capital Programs*, National Defense University, Defense Horizons, January 2011.

[7]   Brown, Winka, and Lee, 2008.

Red Planet Capital (RPC)

In September of 2006, the National Aeronautics and Space Administration (NASA) announced the agency's partnership with a new strategic investment entity named Red Planet Capital. The press release described Red Planet as "a non-profit organization that will establish strategic venture capital for NASA." It was intended to "use venture capital and a NASA investment of approximately $75 million over five years to attract private sector innovators who typically have not done business with the agency," and to be complementary to other investment tools used by NASA for promoting private-sector participation, including the SBIR program.[8] RPC's initial website described intent to invest across sectors that included IT and communications, biomedical support, environmental systems, man-machine systems, smart manufacturing, energy, and advanced materials.[9] RPC made only one investment, in an anti-gravity treadmill company named AlterG.[10] It ceased operations in 2007 when the Office of Management and Budget gave guidance that government-run venture funds would not receive funding in subsequent budget years.[11] This policy apparently did not apply to IQT, which continued to be funded by the Intelligence Community. This policy choice reflected the Bush administration's concern that government-sponsored venture capital projects might displace private funding, and that clear mission-related gaps between government agency and private-sector technology development practices had not been demonstrated for the fledgling NASA and DOE initiatives. Since RPC never advanced past the organization stage, it will not be examined further here.

Rosettex Technology and Ventures Group (RTVG)

In 2002, the National Imagery Management Agency (NIMA) awarded a contract to the newly formed Rosettex Technology and Ventures Group, a joint venture of the Sarnoff Corporation and SRI International. RTVG was formed to address NIMA customer needs in geospatial intelligence, information processing, analysis and management, and digital technology infrastructure.[12] Its business model combined a number of features: R&D services, prototype development and demonstration, system integration, and transition of technology into the commercial marketplace.[13] It involved a wide range of partner organizations performing needs analysis, R&D, product development, and system integration.[14] It worked with NIMA through the National Technology Alliance (NTA), a program that NIMA operated as executive agent for various government agencies,[15] established in 1987 to "foster relationships with critical commercial technology sectors, reduce the barriers that inhibit commercial firms from working

---

[8] D. Steitz and P. Banks, *NASA Forms Partnership with Red Planet Capital, Inc.*, News Release 06-317, September 20, 2006.

[9] Red Planet Capital, *Investment Sectors*, web page, undated.

[10] AlterG, *Learn More About Us*, web page, undated.

[11] Conversation with former senior executive, NASA, January 28, 2012.

[12] Rosettex Technology and Ventures Group, *Rosettex and NTA*, web page, 2004.

[13] Department of Defense, *Annual Report on Cooperative Agreements and Other Transactions Entered into During FY2004 Under 10 USC 2371*, undated b, p. 17.

[14] D. Caterinicchia, "NIMA Tries Venture Capital Route," *Federal Computer Weekly*, March 24, 2002.

[15] T. Benjamin, *Venture Capital Concept Analysis*, Homeland Security Institute, December 2005, p. 23.

directly with the Government, and motivate them to address [Intelligence Community] problems by considering community needs in new product development."[16]

The independent Rosettex Venture Fund (originally projected to total $50 million within ten years) was to be funded by profits from the NIMA-RTVG contract and was to provide "seed and early-stage capital investment to companies with promising solutions to government mission needs."[17] Soon after the formation of RTVG, the U.S. government determined that it could not participate in the Rosettex Venture Fund, and the fund never fully operated.[18]

Why "Other Transaction" Authority Has Been Particularly Important

In all of the GSI initiatives mentioned, an important design element and determinant of success was the statutory authority under which each organization received and disbursed government funds. The use of "Other Transaction" (OT) authority was essential, and it is difficult to understand the more detailed case descriptions below without some understanding of OT authority.

OT authority had its origins in the National Aeronautics and Space Act of 1958, and over the years the statutory authority to engage in OTs has been extended to a number of other agencies, including DoD, the Federal Aviation Administration, the Department of Transportation, the Department of Health and Human Services, and DOE. OT authority is particularly valuable to invoke when the U.S. government needs to engage commercial sources of R&D and prototypes that would not otherwise do business with the government because of the requirements of the FAR. When using OT authorities, there are a number of statutes and government regulations under the FAR that do not apply, and this lends significant flexibility to the contracting agency in the ways that it negotiates and structures agreements with private companies. The excluded statutes and regulations include the Competition in Contracting Act, Truth in Negotiation Act, Contract Disputes Act, and the Procurement Protest System.[19]

DoD OT authority originated with the passing of Public Law (P.L.) 101-189 §251,[20] which authorized DARPA OT authority for R&D projects. Table 3.2 summarizes the evolution of OT authorities for DoD. The initial authorization of OT for DoD had a number of specific restrictions, including that the OT projects not be directly relevant to weapon systems, that project costs be shared equally with other parties, and that this unique authority be used only in instances when standard procurement contracts and grants were not feasible or appropriate.[21] Even in this limited form, OT provided DARPA with the ability to craft R&D agreements with innovative companies that relieved some of the requirements for disclosure, accounting and administration, and assignment of intellectual property rights present in the FAR.

---

[16] Reardon and Scott, 2009, p. 1.

[17] Benjamin, 2005, p. 23.

[18] Conversation with former senior manager of the Rosettex Technology and Ventures Group, December 20, 2012.

[19] D. Sidebottom, *Innovative Contracting Methods*, briefing, April 19, 2010.

[20] Public Law 101-189, National Defense Authorization Act for Fiscal Years 1990 and 1991, Section 251, Allied Cooperative Research and Development, November 29, 1989.

[21] L. E. Halchin, *Other Transaction (OT) Authority*, Congressional Research Service, RL34760, January 27, 2010.

Table 3.2
Evolution of Statutory Basis for Other Transaction Authority

Authorizing Statute	Projects Allowed	Characteristics
P.L. 101-189 §251	Basic, advanced, and applied research, but not building of prototypes	• Cannot be directly relevant to weapons systems • Requires equal cost-sharing with other parties • Only used when standard procurement contracts and grants are not feasible
P.L. 103-160 §845 P.L. 104-201 §804 P.L. 106-398 §803	Can include building of prototypes	• Can be directly relevant to weapon systems • Requires 1/3 cost-sharing OR one nontraditional defense contractor participating, OR exceptional circumstances • Competition only to the maximum extent practicable

SOURCE: RAND analysis.

DARPA subsequently sought more flexibility than P.L. 101-189 afforded, and P.L. 103-160 §845[22] extended OT beyond R&D to include prototypes.[23] A series of changes have been made since section 845 was enacted, so that today OT can be directly relevant to weapon systems; require either one-third cost sharing, one nontraditional defense contractor, or exceptional circumstances; and require competition only to the maximum extent practicable.[24] These requirements leave a fair amount of room for interpretation, particularly regarding the terms "nontraditional defense contractor" and "maximum extent practicable."[25] For example, although a statutory definition of "nontraditional defense contractor" exists in P.L. 103-160, "it is unclear whether, and how, agencies with OT authority verify that a company is a nontraditional contractor."[26]

OT authority has been used or referenced for each of the GSI enterprises examined below.[27] Appendix B includes excerpts from several of the statutes that define OT for DoD, namely P.L. 101-189 §251, P.L. 103-160 §845, P.L. 104-201 §804,[28] and P.L. 106-398 §803.[29]

In the face of persistent urgent operational military needs since 2001, there have been a series of modifications to existing statutes and federal regulations that have allowed the Secretary of Defense to formalize the difference between acquisition paths that address "rapid"

---

[22]   Public Law 103-160, National Defense Authorization Act for Fiscal Year 1994, Section 845, Other Transactions (OTs) for Prototype Projects, November 30, 1993.

[23]   J. Drezner, G. Smith, and I. Lachow, *Assessing the Use of "Other Transactions" Authority for Prototype Projects*, Santa Monica, Calif.: RAND Corporation, DB-375-OSD, 2002.

[24]   R. Dunn, *Injecting New Ideas and New Approaches in Defense Systems: Are "Other Transactions" an Answer?* Naval Postgraduate School, Annual Acquisition Research Conference, May 2009, p. 7.

[25]   G. Fike, "Measuring 'Other Transaction' Authority Performance Versus Traditional Contracting Performance: A Missing Link to Further Acquisition Reform," *The Army Lawyer*, 2009.

[26]   Halchin, 2010, p. 25.

[27]   For more information on OT authorities, see Halchin, 2010.

[28]   Public Law 104-201, National Defense Authorization Act for Fiscal Year 1997, Section 804, Modification of Authority to Carry Out Certain Prototype Projects, September 23, 1996.

[29]   Public Law 106-398, Floyd D. Spence National Defense Authorization Act for Fiscal Year 2001, Section 803, Clarification and Extension of Authority to Carry Out Certain Prototype Projects, October 30, 2000.

as opposed to more "deliberate" acquisitions.[30] These modifications allow the Secretary of Defense to relax provisions of law, policy, directive, and regulation that have to do with establishing requirements, performing R&D, and contracting for urgently needed equipment. These changes are generally referred to as Rapid Acquisition authorities, and they share some of the regulation-relaxation features of OT authorities. However, they are strictly limited to uses that arise from urgent operational need. Therefore, Rapid Acquisition authorities are not suitable for long-term strategic investment initiatives, for which the argument of urgent operational need cannot be made.

Examining Case Details

Having narrowed consideration to three cases, the narrative now recalls the four-actor model from Chapter Two and expands upon it to describe a series of operational functions performed by each actor (Figure 3.1). This model will facilitate discussion of similarities and differences between IQT, RTVG, and OPT.

Figure 3.1
Archetypal Functions and Flows of Value, Private Venture Capital

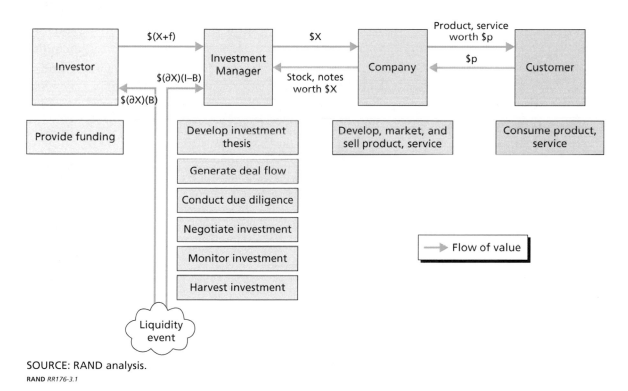

SOURCE: RAND analysis.
RAND RR176-3.1

---

[30] These have included P.L. 107-314 §806c, as amended by P.L. 108-375 §811, and are reflected in such provisions as 6.302-2 of the FAR: Unusual and Compelling Urgency Requirements. A good summary of this subject is contained in Office of the Under Secretary of Defense for Acquisition, Technology, and Logistics, *Fulfillment of Urgent Operational Needs*, Report of the Defense Science Board Task Force, July 2009.

Private Venture Capital Archetype

Each of the GSI cases is an excursion from the private venture capital archetype depicted in Figure 3.1; a precursory description of the archetype will set the stage for the case descriptions. In Figure 3.1, the private investor provides funds ($X+f) to the investment manager.[31] In exchange, the investor expects to receive a large percentage of profits from the investment if and when there is a liquidity event, and the investor and investment manager split this amount based on an agreement made in advance. If all goes well, a multiple of capital "∂" is generated through a liquidity event, and the percentage of profits taken by the investor is "ß." The investment manager, knowing it has $X to put to work, develops an *investment thesis* as the basis for scanning companies participating in market segments of interest.[32] This investment thesis is typically based on analysis of gaps in capabilities among existing offerings in a particular market, or anticipation of development of a new market. It results in the investment manager scanning for technology or business-model solutions to important commercial customer problems.

With an investment thesis in mind, the investment manager must generate *deal flow*,[33] make it known that he or she has capital available, and become well enough established in the network of investors and entrepreneurs that he or she can review multiple products and services that might satisfy the investment thesis.[34] The thesis evolves as the investment manager learns more about existing market participants, competing approaches, and other factors. At some point in this process of scanning the market and testing the thesis, the investment manager finds either that a provisional decision to invest in a particular company is justified, or not. Having chosen a particular company, the investment manager conducts a detailed evaluation (due diligence) of its merits and failings. The due diligence involves many analyses, including evaluations of management experience and ability, company financial performance and projections, existing capital structure and investors, quality of customer relationships, current and proposed business model, size of market, and the like. Sometimes due diligence on a candidate portfolio company will come to an unsatisfactory result, for reasons such as the lack of experience of the management team, or the discovery that the market for the company's product is

---

[31] The investor and investment manager are typically interested in investing in multiple companies, to build a portfolio. This explanation proceeds as if the investor and investment manager are only responsible for the single investment, represented by $X. The "f" represents a fee paid to the investment manager as partial compensation for services provided.

[32] An investment thesis is a proposition that investment objectives will be achieved if and only if investments are made in companies with specific operational and technological characteristics. As an example, consider the challenge associated with making new investments in information system security software. This is a large market, estimated to be greater than $17 billion in 2012, with numerous sub-markets and hundreds of participating companies. New companies enter the market frequently, offering products and services that are often built on new technology. An investor interested in this market typically organizes his or her search based on an investment thesis, such as "the current market is saturated with deep-packet inspection product companies, and will soon demand intrusion-detection capabilities based on real-time analysis of content. I will focus on investing in a promising new provider of this emerging capability."

[33] Deal flow is a stream of investment opportunities originating from a combination of entrepreneurs seeking capital, entrepreneurs being referred by other investors, and entrepreneurs that the investment manager identifies as pursuing a business plan that fits his or her investment thesis.

[34] Here is an example of an investment thesis from the late 1990s that proved true: "Growth in demand for bandwidth over fiber networks is growing at such a rate that there will be sustained strong demand for optical component companies whose products can inexpensively multiplex communication channels for large network infrastructure owners. Large established telecommunication equipment vendors would therefore be willing to pay a substantial premium to acquire companies offering such products."

not as large as originally anticipated. In the former event, the investment manager may choose to evaluate another candidate company under the same investment thesis; in the latter event, he or she may desist from evaluating further candidates because the investment thesis has been disproven.

At some point in the due diligence process, the discussion between the investment manager and the senior leadership of the company turns to the terms on which the investment manager is willing to invest, and conversely the terms on which the company is willing to accept the investment. A negotiation ensues that requires the resolution of many issues, including pre-money valuation, size of investment, type of equity or debt instrument, particulars of corporate governance, and the like. If this process reaches a successful conclusion and the investment is made (an exchange of cash for stock in Figure 3.1), the nature of the relationship between the investment manager and the company changes; the investment manager often joins the company's board of directors and becomes active in advising the company on matters of general and financial management, financing strategy, engineering, product management, sales, marketing, and business development. This relationship can go on for years, until the company has developed its business to the point that it can achieve liquidity for the investor.

OnPoint Technologies

Of the three cases, OPT is the shortest departure from the private venture capital archetype (see Figure 3.2). In 2003, MilCom Technologies won a competition among 30 bidders to manage

Figure 3.2
Functions and Flows of Value, OnPoint Technologies

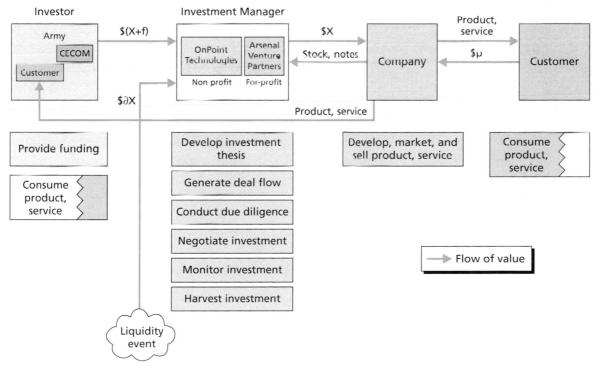

SOURCE: RAND analysis.
NOTE: In this figure, the split shading associated with the customer box highlights that consumption of portfolio company products and services are shared between Army customers and commercial customers.
RAND RR176-3.2

a new Army-sponsored GSI initiative that was originally named the Army Venture Capital Investment Corporation (VCIC). House of Representatives Report 107-298 states that the intent of the House Committee on Appropriations was to model the VCIC on IQT. MILCOM Venture Partners was formed subsequently, and it managed both VCIC (later named OPT) and MilCom Technologies. This arrangement was necessary because VCIC was required by statute to be non-profit, and therefore the for-profit MilCom Technologies was not a suitable long-term host. MILCOM Venture Partners was itself later renamed Arsenal Venture Partners.

In this case, the U.S. Army is the strategic investor and its Communications Electronics Command (CECOM) serves as the principal interface with OPT.[35] Along with providing contractual and administrative support, CECOM has helped OPT to identify priorities for innovations needed to improve mobile power solutions for soldiers, and to maintain ties between OPT and Army customers for solutions offered by the companies. The Army provides funds ($X+f) for investments in portfolio companies ($X) and for expenses at OPT and AVP (f). OPT does not have financial resources to pay for prototype developments for prospective Army customers (unlike IQT, as described below). Proceeds from liquidity events are distributed somewhat differently that in the private venture capital archetype (Figure 3.1). All funds flow back to OPT, with a portion being paid to AVP as part of their compensation, while the rest is retained by OPT for reinvestment.

CECOM utilized OT authority in structuring both the contract between OPT and the Army and the management agreement between OPT and AVP. The use of OT authorities made it possible for the Army to structure these agreements using procedures, terms, and conditions that are closer to standard commercial practice than would have been possible under the FAR. OPT does not provide non-equity funding to portfolio companies for the development of Army mission-specific prototypes.

As in the archetypal private venture capital reference of Figure 3.1, responsibility for all of the investment functions (highlighted in green) resides predominantly with the investment manager, AVP. Their development of investment theses focuses on technology and market opportunities pertaining to mobile power solutions. However, one very important difference is that, in this case, the investment manager must not only attend to the quality of *commercial* market opportunities for potential portfolio companies, but also to the quality of the *Army mission-oriented* market opportunities. There has been no apparent need for investment functions to be shared outside the investment manager; the sector-specific nature of the portable power investing focus has allowed the principals at AVP to become sufficiently expert that they can sustain adequate deal flow, conduct thorough due diligence, and monitor investments effectively.

OPT's employee incentive system is similar to that of a typical venture capital firm.[36] However, the Army also built an incentive into their agreement with OPT that flows through to OPT's management agreement with AVP, awarding compensation in part based on the degree of adoption of portfolio company solutions by Army customers. Although OPT does not provide direct financial support for portfolio companies to conduct prototype developments, the anticipated willingness of the Army to buy company solutions strongly influences how AVP selects and manages investments. Managers at AVP expend significant effort creating prospective demand from Army customers before making investments.

---

[35] Benjamin, 2005, p. 30.

[36] Conversations with senior manager of Arsenal Venture Partners, July 19, 2012, and June 20, 2013.

OPT received limited funding appropriated specifically for its use; the majority of its funding came in reallocation from other parts of the Army's R&D budget. Section 8150 of the fiscal year 2002 DoD Appropriations Act earmarked the initial funding to establish VCIC,[37] and the Army subsequently reallocated $25.4 million, $12.6 million, and $10.0 million in fiscal years 2002, 2003, and 2004, respectively. Decisions on which projects to invest in are made by OPT, with focus on innovative technologies of interest to the U.S. Army. As the OPT website explains, "The U.S. Army is now developing concepts and seeking technologies that will vastly increase the battlefield effectiveness of the individual, foot-mobile soldier. These new concepts will require power and energy sources that are significantly improved over devices currently available." OPT has invested steadily through the years since its creation, and, as of May 2013, lists a portfolio of 12 companies on its website.[38] These include Atraverda (battery electrodes), Nanosolar (technology for printing solar cells on flexible substrates), and Power-Precise solutions (battery management devices). OPT management believes that long focus on mobile power solutions has allowed them to build a unique breadth and depth of understanding of market offerings from innovative private companies—understanding that the Army would have had difficulty building through other means.

OPT defines the terms of individual investments in private negotiations with the companies and their other investors. Because of OPT's relationship with the U.S. Army, it is able to assist portfolio companies in managing information rights and other requirements under the FAR. While OT authorities were used to form OPT itself, any volume of the Army's purchase of products and services from OPT companies is conducted under the FAR.

In-Q-Tel

IQT differs from the private venture capital archetype of Figure 3.1 in a number of important respects (Figure 3.3). As with OPT, the investor is no longer at arm's length from the company, but has become an important customer. But there is now also an explicit exchange of value in addition to the equity investment ($dX in Figure 3.3), namely a work program contract worth $(1-d)X to deliver a prototype solution addressing an investor mission need. This prototype is adapted for the purpose from the portfolio company's commercial product or service. The delivery of the prototype is managed by IQT with specific performance and information rights being defined using OT-like authorities; the specifications for the prototype are determined in consultation with the CIA customer, but the contract is between the company and IQT. If the prototype activity is judged by the CIA customer to be successful, then further purchases of the product are the financial responsibility of CIA, not IQT, and are managed through conventional CIA acquisition processes. With the addition of the mission-need component as a determinant of the desirability of an investment, it becomes important for the investment manager to know and be able to translate investor needs into specific performance by the company. The investment management functions now also explicitly include the need to manage both financial investments and prototype development work programs.

As indicated by the mixed-shade boxes in Figure 3.3, some functions that were performed in the OPT case exclusively by the investment manager are now shared with the investor interface organization, in this case the In-Q-Tel Interface Center (QIC). The QIC was set up

---

[37] Office of the Assistant Secretary of the Army for Financial Management and Comptroller, *Source of Funds for Army Use (Other than Typical Army Appropriations)*, Resource Analysis and Business Practices, SAFM-RB, March 2005, p.14.

[38] OnPoint Technologies, *Portfolio*, web page, undated.

Figure 3.3
Functions and Flows of Value, In-Q-Tel

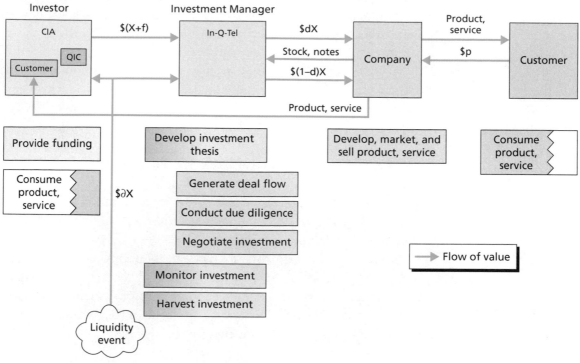

SOURCE: RAND analysis.
RAND RR176-3.3

to facilitate the discovery and communication of mission priorities to IQT staff (the "problem set"), market IQT capabilities within the CIA, and to provide IQT with governance and administrative support. The process of developing investment theses is now shown as shared between investor and investment manager; the QIC plays an important role in articulating government market needs, while the technical and financial specialists within IQT translate this understanding into appropriate investment theses based on their knowledge of related commercial market needs. This IQT staff (a mix of technical and financial transaction specialists) then generate appropriate deal flow, conduct due diligence, and negotiate investments and work programs. The process of monitoring investments is shared among the CIA customer, the QIC, and IQT, and success now has two components: transition of prototypes from the work program into operational use by CIA customers, and the private capital markets process of generating investment liquidity.[39]

IQT exercises influence on portfolio companies through both the investment and prototype contract mechanisms. The investment provides IQT visibility into the entirety of the company's plans and performance, and IQT staff participate as observers to the company board of directors. The work program sometimes calls for the adaptation of product features or functions that are already in the company product plan at the time of the IQT investment. But sometimes the work program provides the company the opportunity and incentive to either

---

[39] Conversation with former In-Q-Tel staff technical specialist, October 31, 2012.

accelerate introduction of features or functions, or to introduce new features or functions that may also have commercial market value.

Since IQT is non-profit, the technical and transactions specialists who perform the investment management functions are not compensated in the same way as their peers in private capital, or as the investment managers in AVP (see OPT case). Employees receive a base salary and are eligible for an annual cash bonus[40] based on factors that include uptake of prototype solutions by CIA customers. But they do not share profits from successful investments to the same degree as their peers in private venture capital funds. IQT once had a mandatory participation "Long-term Incentive Compensation Fund" that was invested alongside IQT funds into portfolio companies, and the associated returns were distributed to participating employees. But the new IQT CEO removed this compensation element in 2006, as part of an initiative to strengthen incentives for staff to focus more on uptake of solutions by the Intelligence Community, and less on financial returns. In spite of this shift in financial performance–based compensation, the technical and financial transaction specialists at IQT continue to be viewed by private capital peers as having valuable market understanding and investment management capability.

When IQT was created, the CIA General Counsel worked to develop a contract vehicle that provided the new non-profit adequate operating flexibility to effectively pursue both elements of its business model: investments and prototype work programs. In doing so, the CIA could have relied on the broad authorities of Section 8 of the CIA Act of 1949,[41] but chose instead to emulate elements of the DARPA organizational model as based on OT authority. As Yannuzzi describes it:

> Using a DARPA model OT agreement as a guide, the [CIA] designed a five-year Charter Agreement that describes the general framework for its relationship with In-Q-Tel, sets forth general policies, and establishes the terms and conditions that will apply to future contracts.[42]

Thirteen years after its creation, IQT continues to refer to the OT authority precedent, and these OT authority practices served as an example for other GSI initiatives, such as those of OPT and RPC.

CIA funding of IQT has been relatively stable through the life of the organization; during fiscal years 2002–2006, CIA funded IQT at approximately $40 million per year, and in each fiscal year money was expended over a three-year period. The result has been a fairly steady expenditure profile. Decisions about which companies to engage were made by IQT staff, relying on guidance and consultation with government employees among the Intelligence Com-

---

[40] Business Executives for National Security, 2001.

[41] Authorizing the CIA to spend funds "for purposes necessary to carry out its functions … notwithstanding any other provision of law" (L. B. Snider, *The Agency and the Hill: CIA's Relationship with Congress, 1946–2004*, Washington, D.C.: The Center for the Study of Intelligence, Central Intelligence Agency, 2008, p. 154).

[42] R. Yannuzzi, "In-Q-Tel: A New Partnership Between the CIA and the Private Sector," *Defense Intelligence Journal*, Winter 2000.

munity customer base. In 2012, IQT was reported to have received $56 million in government support.[43]

The terms of investment by IQT and private companies are determined by IQT through negotiation with the companies and their other investors. For actual purchases of stock and debt, IQT typically assumes the terms negotiated by the private investors. Instead of paying cash for stock or notes, IQT instead often receives stock warrants for a number of shares equivalent to 20 percent of the value of the contract to provide a technology prototype to the Intelligence Community customer. The terms of this contract are negotiated by IQT consistent with OT authority, being mindful of the eventual intent that successful prototypes will be purchased directly by CIA. IQT typically does not have the option or authority to initiate a liquidity event through special shareholder or board of directors voting rights or redemption rights. These kinds of decisions are determined by shareholder agreements and decisions by the portfolio company board of directors, to which IQT has observer rights—meaning that a representative of IQT has the right to attend and participate in deliberations of the company's board of directors. This is a valuable mechanism for staying informed about the entirety of a company's business in a way that is not possible through conventional acquisition practices. If the liquidity event associated with any company results in profits to IQT, these are allocated per a memorandum of understanding between IQT and the CIA to be split evenly between additional IQT investing activity and strategic information technology initiatives defined by the CIA.[44]

IQT has invested in over 200 companies since its beginnings in 1999. In technology areas of importance to its government customers, IQT often engages with a range of companies operating in any particular market. Because it is known to be active in providing contracts to develop prototypes for potential scale purchase, IQT has the opportunity to evaluate the products and business plans of many potential companies, including competitors in the same market. This allows IQT to build broad knowledge of private companies and their potential solutions, and to engage quickly and broadly with companies creating technologies of emerging importance. One example of this in recent years is the investment portfolio IQT has built in big data and advanced analytics companies. IQT has invested throughout the market that provides such solutions and services, from Cloudera, Inc., and 10Gen, Inc., who provide analytics and operational infrastructure to Palantir, Inc., and Recorded Future, Inc., which provide analytics applications and visualization solutions. The information gathered in these relationships and prototype activities helps the Intelligence Community to exploit emerging data management capabilities for intelligence missions.

Rosettex Technology and Ventures Group

The arrangement of functions and flows of value for RTVG (Figure 3.4) is a further departure from the private venture capital archetype than either OPT or IQT. Although there are still parallels between the cases (e.g., the importance of the respective government customers), there are also notable differences. For example, the singular "company" of the other cases is now a partner team comprising multiple companies. RTVG managed the workflow of numerous partner companies, ranging from small single-product firms to large and established system-

---

[43] National Public Radio, *In-Q-Tel: The CIA's Tax-Funded Player in Silicon Valley*, All Tech Considered transcript, July 16, 2012.

[44] Mohlzahn, 2003.

Figure 3.4
Functions and Flows of Value, Rosettex Technology and Ventures Group

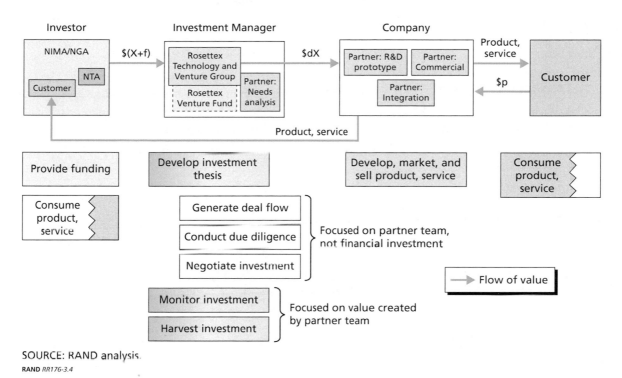

SOURCE: RAND analysis.
RAND RR176-3.4

integration firms. There is also a difference in how customer interface purposes were served. NTA was not so important to fostering customer relationships as were either the QIC or CECOM.[45] The organization of investment manager functions has some features that are similar to the other cases (e.g., the sharing of investment thesis development between investor and investment manager), and some that are different (e.g., investment theses having to do with partner company capability rather than with suitability for equity investment). A related point of difference is that due diligence and monitoring was focused on partner team formation and operations rather than on selection and management of equity investments. The final RTVG project portfolio described the business model this way:

> . . . Rosettex business model with all the elements required to make this happen, namely: a contract mechanism designed for rapid procurement actions, an organizational structure designed to streamline management, a team with clearly defined roles to exploit their expertise, and an experienced management team ready and able to work in close cooperation with its Government clients and industry partners to facilitate access to best-of-class resources and provide independent project oversight.[46]

One of the novel features of the RTVG initiative was a venture fund component that was intended to invest profits from the RTVG joint venture into innovative companies that would participate in the partner network. The Rosettex Venture Fund was put into the original pro-

---

[45] Conversation with former senior manager of the Rosettex Technology and Ventures Group, December 20, 2012.

[46] Reardon and Scott, 2009, p. 1.

posal for RTVG, and was referred to as an "on-call" fund, with the intent that a portion of the profits on the core contract for the joint venture would be available for investments that would be chosen by a government-convened board, from a pool of possibilities presented by the RTVG. Because of difficulties in identifying legal authorities under which the joint venture could take investment direction from a government-sponsored entity, the Rosettex Venture Fund never fully operated. Thus, RTVG's incentive to perform its partner team–building and coordination functions never had a capital gains component. Consequently, a smaller number of participating companies fit the early-stage venture-backed profile that dominates the IQT and OPT portfolios. A larger number were established companies with various expertise. The Rosettex partner teams were designed to be more vertically integrated than any combination of companies in the OPT or IQT cases. One of the partner team announcements from 2002 explains further:

> Rosettex has assembled a team of 64 partners representing major technology consulting firms, established and new commercial companies, independent research institutes, premier academic institutions and government contractors. Access to the vast array of research and technologies represented by these organizations will allow the Rosettex team to accelerate the transition of a wide range of technologies from the laboratory to the commercial marketplace to address the government users' needs.[47]

To facilitate rapid procurement actions, RTVG utilized OT prototype authorities both for the joint venture and for individual partner or team projects.[48] Many examples of projects pursued by RTVG are available in the final project portfolio report of NTA. These ranged from studies and assessments, such as a commercial technology trends analysis conducted by Gartner for NGA, to technology prototype activities, such as a demonstration of hyperspectral/panchromatic data fusion by Carnegie Mellon University and Leica Geosystems, again for NGA. Although the majority of projects were directed at NGA customers, RTVG also managed projects for the Office of the Director of National Intelligence, Special Operations Command, and for JIEDDO.

In 2002, NIMA obligated $100 million to RTVG under a contract to expend the money over a period of five years. The obligated funds were expended by the private joint venture in a fashion that resembles task-order management under an indefinite delivery/indefinite quantity (IDIQ) contract. The IDIQ approach to acquisition allows the government to place incremental orders for supplies or services against a larger omnibus contract. RTVG was able to commit and expend funds with greater flexibility than in a typical IDIQ, because from 2002–2006 the RTVG joint venture was itself deemed to be a nontraditional defense contractor, and thus qualified to operate under OT authority. The choice of projects and participants was strongly guided by the RTVG joint venture. Although government approval was required, this approval was usually awarded rapidly. An RTVG program manager and government program manager worked closely together. At one point in the 2004–2006 period, there were over 100 active projects and 100 partner companies in the participant network.

---

[47] M. Seebold, *Saffron Technology, Part of the Rosettex Technology and Ventures Group, Wins Major Award from US Government's National Imagery and Mapping Agency (NIMA)*, Saffron Technology press release, May 2, 2002.

[48] See Department of Defense, *Annual Report on Cooperative Agreements and Other Transactions Entered into During FY2003 Under 10 USC 2371*, undated a, p. 16–21.

In 2006, the RTVG private joint venture was determined to no longer qualify as a non-traditional defense contractor under OT authorities, at least in part out of concern within Congress and the Office of the Secretary of Defense about the use of OT authority by the Army and Boeing to utilize an 'other transaction agreement' for system development and demonstration of the Future Combat Systems program.[49] Consequently, after 2006, the process of defining and choosing projects required more justification and administration, with every project having to stand on its own as eligible under OT authorities. This greater degree of oversight reduced the rapidity with which RTVG was able to initiate and conduct projects, and contributed to a decision on the part of NGA not to renew the core contract in 2009, at which time RTVG ceased operations.

Conclusion

Each of these GSI initiatives found a balance among financial incentives that allowed for a level of engagement with private companies that suited investor agency needs. They each used a different balance, with OPT placing the most emphasis on investment and RTVG the least. They each made use of OT (or OT-like, in the case of IQT) authority to tailor relationships so as to encourage innovative companies to participate in their government mission-oriented activities. A number of themes or lessons emerge from the three cases, and they are presented in Chapter Five. Before that summation, Chapter Four will examine the GSI choice among incentives through economic modeling.

---

[49] Halchin, 2010, p. 30.

CHAPTER FOUR

Economic Framework for Innovation Incentives

The case studies of Chapter Three discuss interactions among the government, the investment manager within the GSI initiative, the portfolio company, and the customer. We can identify several key features that distinguish the problem of optimal GSI as different from that of the private venture capital archetype. First, a government strategic investor cares not only about maximizing return on equity but also about generating prototypes that are specifically useful to the government. Second, the probability of a successful sale of the ensuing product will differ depending on whether the final customer is the private sector or the government, and this difference is of greater interest to GSI because of the first factor. A higher probability of sale would, to a certain degree, attenuate demand risk (the risk that an innovation is created but no customers surface to purchase the final product). Third, as we observed in the case studies in Chapter Three, GSIs routinely employ a mix of incentives by allocating funds either to an equity investment that supports R&D within a company or to a work program contract for developing government-specific prototypes. OPT most closely followed the traditional venture capital archetype by allocating all funds to equity investment. IQT exercised influence through both the investment and contract mechanisms, while RTVG chose to place all of its emphasis on contract incentives. In this chapter, we describe an economic model that incorporates these features to draw generalizable insights regarding the circumstances under which optimal outcomes may be achieved.

Although numerous incentives may influence the overall development of innovations, this chapter focuses solely on the incentives relevant to the portfolio companies. The economic model accepts that a GSI initiative already exists, that the government has determined the level of resources to commit through the initiative, and that the initiative has determined which portfolio company will receive these resources. The model includes no incentives based on competition between portfolio companies for GSI resources, and it does not address the structure of employee compensation incentives within the GSI. In fact, the work of the GSI investment manager in choosing portfolio companies is complete in the model—the investment thesis is assumed to be mature and stable. Outside the scope of the model is the structure of incentives associated with the government's potential purchase of products that flow from prototypes that the GSI funds—for example, purchase guarantees, financing arrangements, or performance bonuses. The strict purpose of the economic model is informing the GSI's allocation decision of how aggressively to push portfolio companies to develop government-relevant prototypes.

We propose a simplified two-stage game-theoretic model that addresses the relationship between GSI and a single innovating firm. In the first stage, given a fixed budget, the GSI makes an allocation choice. Funds may be allocated to the firm in the form of a general investment to support R&D, or in the form of a contract for development of a government-specific prototype. This is depicted in Figure 4.1 as the choice between $dX and $(1-d)X. When the firm invests in R&D, there is a fixed probability (λ) that each dollar spent will generate an innovation. This construct serves as an abstract representation of the process by which R&D investment leads to a total number of innovations. We assume that any innovation can become a prototype either developed specifically for the private sector or specifically for the government. However, certain innovations may be more naturally suited to development for government use, while others may be more naturally suited to development for private-sector customers. The model assumes the proportion of innovations naturally skewed to the government (γ) is a fixed value, and because the investment thesis is stable, the GSI's investment manager has no further interest in influencing this value. If a prototype is developed for a sector to which the innovation is *not* naturally suited, a *rechanneling cost* is incurred. The contract amount spent on $(1-d)X may be interpreted as a subsidy to offset the costs of rechanneling innovations naturally suited to the private sector into government-specific prototypes.

In the second stage of the model, the firm chooses the proportion of innovations to develop into prototypes for the government (α_G) and the private sector (α_P). Although the firm can pay to either rechannel innovations from the government to the private sector or from the private sector to the government, the $(1-d)X contract funds are to be spent only on rechanneling to government-specific prototype development. From the point of view of the GSI, this is desirable because it encourages government-specific prototypes over private sector–specific prototypes from the firm. Regardless of whether the prototype is government-specific or private sector–specific, the firm will attempt to sell all its prototypes to both sectors. Successful

Figure 4.1
Summary of Two-Stage Economic Model

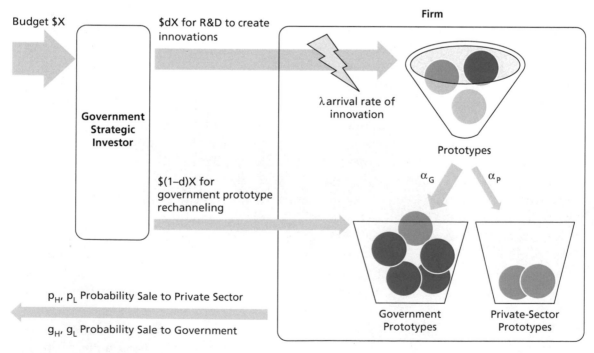

SOURCE: RAND analysis.

sales to each sector can be associated with probabilities of sale that depend on the intended development sector and the sector of actual sale, as shown in Table 4.1. For example, we assume the probability of actual sales to the government is high when the prototype is developed for the government sector (g_H) and low when the prototype is developed for the private sector (g_L). The difference between these two probabilities ($\Delta g = g_H - g_L$) can be interpreted as the government sensitivity to the specificity of the prototype. Likewise, the difference in private-sector probabilities ($\Delta p = p_H - p_L$) is the sensitivity of the private-sector customer. The exogenous probabilities of sale take into account the size of the government and private-sector markets; for example, when the government market is large, the probability of sale to the government will also be large. Total sales depend on the probabilities of sale and the firm's sector-specific development choice in this stage. When the firm dedicates more effort to developing government-specific prototypes, total sales to the government will increase and total sales to the private sector will decrease.

Success for the GSI is measured in terms of the GSI's utility. As mentioned previously, the utility function for GSI differs from that of a traditional venture capital entity. A traditional venture capital firm is concerned solely with financial returns. The utility of a GSI enterprise is instead a function of both financial returns and the firm's sales to the government, which serves as a proxy for government-specific innovation and speaks to the mission of the GSI.

Therefore, there are two possible drivers that would influence the firm's decision to channel development of new technologies toward *government-specific* applications:

1. Pickiness: We define pickiness as the relative sensitivity to the specificity of the prototype development. If the private-sector customer is pickier (i.e., more sensitive) than the government customer ($\Delta p > \Delta g$), the firm will choose to rechannel innovations to the private sector in order to maximize overall sales. Likewise, if the government is pickier ($\Delta g > \Delta p$), innovations will be rechanneled to the government. We assume the probabilities of sale and the degree of pickiness within the market are exogenously determined. This reflects the fact that, in the short term, neither private-sector sensitivity (Δp) nor government sensitivity (Δg) can be easily manipulated in a credible fashion.

2. Subsidy for government prototype rechanneling: These are the $\$(1-d)X$ funds the GSI allocates in the first stage of the model (see Figure 4.1). While the degree of pickiness can drive rechanneling to either sector and is outside the control of the GSI, increasing the share of funds designated for government rechanneling ($1-d$) is a possible lever for the GSI to induce an increase in the quantity of innovation developed for the government sector. A trade-off immediately apparent from the structure of the model is that when the investment portion of funding increases, the total number of prototypes will

Table 4.1
Probabilities of Sale

Intended Development	Actual Sale	
	Government	Private Sector
Government	g_H	p_L
Private Sector	g_L	p_H

SOURCE: RAND analysis.

increase, but the subsidy per government-specific prototype will decrease, which results in fewer government-specific prototypes. On the other hand, when the investment portion decreases, the total number of prototypes will also decrease until there are no longer enough innovations to develop into government-specific prototypes.

The economic model speaks directly to the duality between the two drivers. It will inform under what conditions the GSI can and should exert influence on the firm by aggressively pushing its portfolio company to provide government-specific prototypes with $(1-d)X$. This hybrid innovation-generation model is a game-theoretic model—a dynamic game of complete information. The GSI's choice in the first stage affects the firm's choice in the second stage, and, reasoning backwards, knowing the firm's strategy will determine the GSI's best course of action. Appendix A presents the details of using backward induction to solve for the equilibrium values of the choices made by the GSI (d^*) and the firm (e.g., α_P^* and α_G^*)

Government and Private Sector Choices

Before delving into the solution of the model, we will first clarify the definition of *rechanneling* using Figure 4.2. The length of each bar represents the total number of innovations generated by dX, and the division of colors illustrates the firm's hypothetical choice of the proportion of total innovations to develop into prototypes for each sector. In the top bar of Figure 4.2, the private sector is more sensitive, or *pickier*, than the government, so we would expect some degree of rechanneling to the private sector—the proportion to be developed for the private sector is greater than the proportion that is naturally suited to the private sector ($\alpha_P > 1-\gamma$). The bottom bar illustrates an example of rechanneling to the government. Proposition 1 in Appendix A proves that, in equilibrium, no innovations will be wasted, meaning that every

Figure 4.2
Allocation of Prototype Development

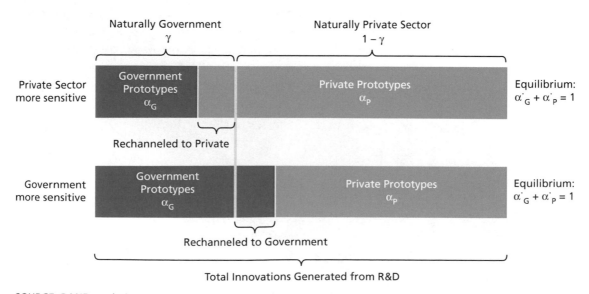

SOURCE: RAND analysis.
RAND RR176-4.2

innovation will be developed into a prototype for either one sector or the other ($\alpha_P^* + \alpha_G^* = 1$). To calculate the total quantity of innovations developed for each sector, one would multiply the corresponding fraction of the bar by the total number of innovations generated. Finally, we make note of a subtle point: The length of the bar, which represents total quantity of innovation generated from the dX investment, changes depending on the GSI's choice of d^*. This is because we define *innovation* very specifically in the model as the probabilistic result of the R&D investment and the arrival rate (λ). The development life cycle is as follows: (1) innovation, (2) prototype, and (3) final product sale.

We will now present the equilibrium solutions and intuition for the GSI's choice of d^*, the firm's prototype development choice (α_G^*, α_P^*), and how they combine to provide an overall picture of budget expenditure. In the upcoming figures, results will be arranged on the x-axis across a continuum of pickiness. On the far right of interval i, the private sector is much pickier than the government (i.e., most sensitive to the specificity of the prototype development). At the origin, both sectors are equally picky ($\Delta p = \Delta g$), and as we move further to the left into intervals iii and iv, the government becomes pickier than the private sector. Although pickiness varies continuously along the horizontal axis, we find that the optimal choices and outcomes change in discrete amounts between intervals.[1]

Figure 4.3 shows the GSI's optimal share of funds for investment (d^*) on the y-axis, and we can also easily infer the optimal share of funds for government prototype rechanneling ($1-d^*$). In intervals i and ii, the government is not picky relative to the private sector, so many private-sector innovations would also appeal to the government customer. Consequently, the optimal policy for the GSI is to provide no subsidy for rechanneling and allocate the entire

Figure 4.3
Optimal Choice d*

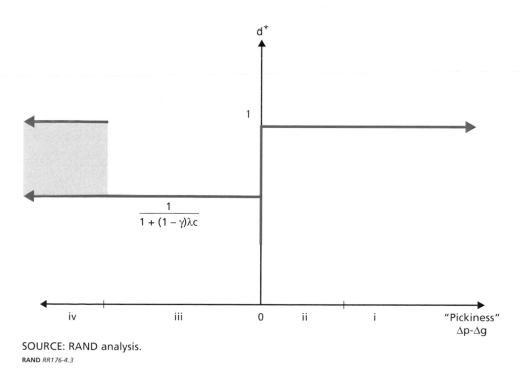

SOURCE: RAND analysis.
RAND RR176-4.3

---

[1]   Intervals i, ii, iii, and iv correspond to Proposition 4 (i), 4 (iii), 4 (ix), and 4 (x) in Appendix A, respectively.

budget to investment to maximize overall innovation ($d^* = 1$). In interval iii, the government customer is pickier than the private-sector customer, and the GSI will elect to subsidize rechanneling toward government-specific prototype development ($d^* < 1$). It does so knowing that allocating part of the budget toward the rechanneling subsidy will deplete the funds available for R&D more generally, resulting in a loss of total innovation but a gain in government-specific prototypes. This behavior is rooted in the assumption that the GSI has a *significant* interest in generating prototypes that are specifically useful to the government in addition to an interest in maximizing the return on equity.[2] For any point on the x-axis within interval iv, the GSI is indifferent among the choices in the vertical range of the shaded box. The intuition is that, upon crossing into interval iv, the government is so much pickier than the private sector that even if the GSI were to allocate no funds toward rechanneling ($d = 1$), the firm would rechannel innovations to the government on its own and cover the cost by drawing on the general fund ($dX). The vertical range of the shaded box represents the share of the budget that would be designated for rechanneling regardless of whether the cost is covered by drawing from general funds ($dX) or contract/subsidy funds ($[1-d]X).

Given our expression for d^* in intervals iii and iv, we can represent the optimal share of funds reserved for rechanneling ($1-d^*$) as a function of the share of innovation naturally suited for government application (γ), the probability of innovation (λ), and the unit cost of rechanneling innovation (c). We find that when the share of natural government innovation increases, the share reserved for rechanneling decreases because there are fewer opportunities for rechanneling. An increase in the probability of innovation affords the GSI more freedom to shift funds toward rechanneling without sacrificing overall innovation. Finally, when the cost of rechanneling innovation increases, the share reserved for rechanneling must increase to induce firms to produce government-specific prototypes. However, if the cost of rechanneling is excessively high,[3] the optimal policy for the GSI changes to where the GSI never chooses to subsidize rechanneling, as seen in Figure 4.4. For the rest of this section, we will assume rechanneling costs are not excessively high.

The trade-off between increasing government-specific rechanneling at the expense of sapping overall R&D and total innovation would be resolved differently if the GSI were to act like a traditional venture capital investor interested only in the return on equity and not in generating government-specific prototypes *per se*. Such a case is illustrated in Figure 4.5, and we can see that such a GSI would behave differently in interval iii by choosing not to subsidize rechanneling. For the rest of this section, we assume that this is not the case and that the GSI does have a significant interest in generating government-specific prototypes.

Given the GSI's choice of the share of funds allocated to the general fund (d^*) versus rechanneling ($1-d^*$), the firm will choose the share of innovations to develop into a prototype geared toward the government (α_G^*) versus the private sector (α_P^*). Recalling one of the lessons of Figure 4.2, namely that the shares α_G^* and α_P^* would sum to one in equilibrium, we can see in Figure 4.6 that when the private sector is much pickier (interval i), the firm will develop every innovation for the private sector. As the effect of pickiness subsides, moving into interval ii, the firm will develop innovations naturally suited for government into government-specific prototypes and innovations naturally suited for the private sector into private-sector pro-

---

[2]   In Appendix A, we provide formal definitions for conditions where the GSI cares about government sales.

[3]   Excessively high is defined as $c > \frac{1}{\gamma\lambda}$.

Figure 4.4
Optimal Choice of d* When Cost Is Excessively High

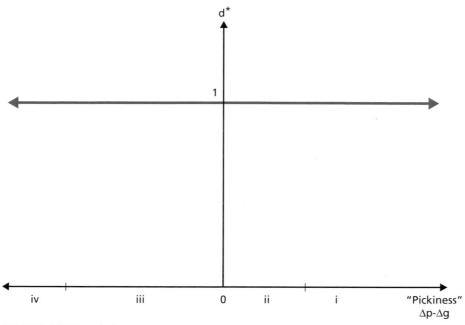

SOURCE: RAND analysis.
RAND RR176-4.4

Figure 4.5
Optimal Choice of d* for GSI Primarily Interested in Equity Return

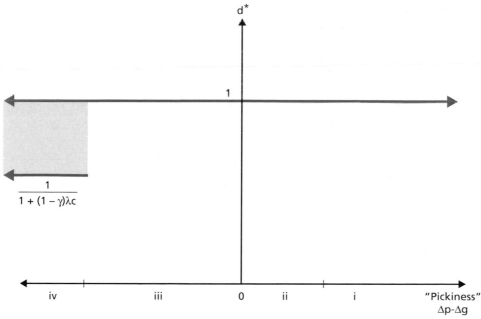

SOURCE: RAND analysis.
RAND RR176-4.5

Figure 4.6
Firm's Prototype Development Choice in Equilibrium

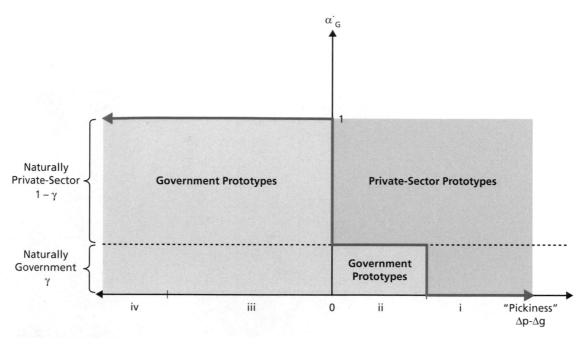

SOURCE: RAND analysis.
RAND RR176-4.6

totypes; that is, no rechanneling occurs. Interval iii in Figure 4.6 reflects the effect of the GSI's choice in Figure 4.3 to subsidize government rechanneling, causing the firm to develop all innovations into government-specific prototypes. In interval iv, both the rechanneling subsidy and the extreme pickiness of the government customer contribute to the firm's choice. The knife-edge results we see here can be attributed to the linearity of our economic model.

Figure 4.7 shows the total quantity of innovations rechanneled, which is defined as the difference between the chosen amount of prototype development (as seen in Figure 4.6) and the natural amount without the effect of rechanneling. On the right side of the vertical axis, rechanneling is oriented entirely toward the private sector, while on the left side, rechanneling is oriented entirely toward the government. On both sides, rechanneling occurs due to pickiness; however, additional rechanneling in interval iii is driven by the GSI's rechanneling subsidy ($[1-d]X). This difference in the level of rechanneling is due to the share of innovation that is naturally suited to government application (γ). If all innovations were equally well suited for either sector (i.e., $\gamma = 1/2$), the heights of the segments on the far left and far right would be equal. Figure 4.7 assumes a minority of innovations is naturally suited for government application ($\gamma < 1/2$); since more government rechanneling (and less private-sector rechanneling) is required, the segment on the left is higher than the segment on the right.

Because the firm can attempt to sell any prototype to both government and private-sector customers (with varying degrees of success due to the probabilities of sale), one line of reasoning might suggest that as the government becomes extremely picky (moving toward the left side of the horizontal axis), the firm's best strategy may be to simply abandon efforts to cater to the government and rather focus more effort in pursuing the less picky private customer. We will now dispel this reasoning given the assumptions we have made so far. Total sales to the govern-

Figure 4.7
Total Quantity of Innovations Rechanneled

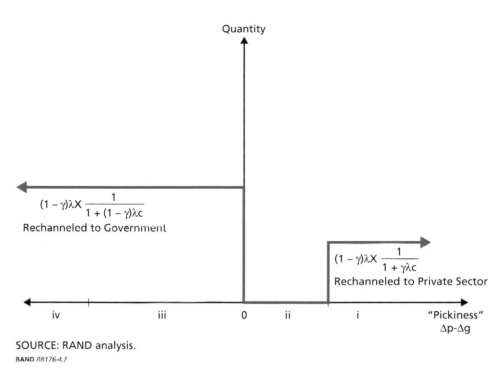

SOURCE: RAND analysis.
RAND RR176-4.7

ment can be calculated by multiplying $\alpha_p g_L + \alpha_G g_H$ by the total number of innovations, and total sales to the private sector can be calculated by multiplying $\alpha_p p_H + \alpha_G p_L$ by the total number of innovations. Figure 4.8 shows that the ratio of government sales to private sales increases as the government becomes pickier. In fact, there is no danger of the firm abandoning the government customer, and government pickiness increases the degree to which firms produce products that are useful for the government.

We will now combine the equilibrium solutions for the GSI and the firm to provide an overall picture and summary of the results, which we depict in Figure 4.9. The total budget available to the GSI (X) is the height of the graph, and for each interval, we will describe the budget expenditure in equilibrium.

- Interval i: Due to the extreme pickiness of the private sector, the firm develops every innovation into a prototype for the private sector. It does so in an effort to maximize sales. The firm covers the cost of rechanneling innovations to the private sector by drawing on general funds ($dX). This drawing of funds is represented by the red box. Covering the cost of rechanneling in this way depletes funds that would otherwise be used for R&D. In addition, a certain fraction of R&D (1-λ) is unsuccessful (grey box).
- Interval ii: At this point along the continuum, the level of pickiness is not sufficient to motivate any rechanneling, and the entire budget (X) is spent on R&D. Of the successful innovations generated, prototype development follows the direction for which the innovation is naturally suited.
- Interval iii: The GSI's equilibrium choice is to subsidize rechanneling to the government (yellow box). The firm responds by using the subsidy to develop every innovation into a prototype for the government.

Figure 4.8
Ratio of Government/Private Total Sales

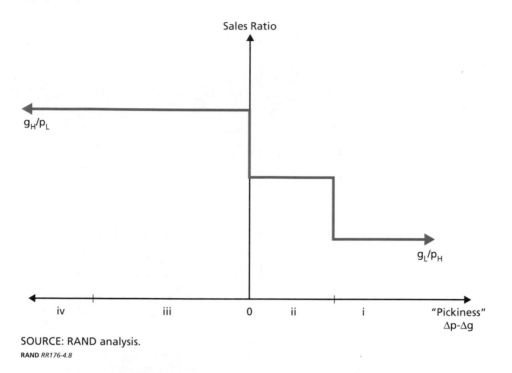

SOURCE: RAND analysis.
RAND RR176-4.8

- Interval iv: Government pickiness is great enough that the firm develops every innovation into a government-specific prototype—even when no subsidy is provided. The GSI need not designate funds for government rechanneling because the firm will cover the rechanneling cost by drawing on the general fund ($dX) anyway.

We define the *influence* exerted by the GSI on the firm as the increase in innovations developed for government customers that results from the rechanneling subsidy ($[1-d]X). The blue line in Figure 4.9 represents the GSI's optimal choice of d*. To identify the influence effect, we construct a counterfactual GSI policy in which no contract for rechanneling is offered under any circumstances (1-d = 0). Solving for the firm's choice given the counterfactual GSI policy results in Figure 4.10. Comparing Figures 4.9 and 4.10, we can see a difference in outcomes in interval iii. Without a rechanneling subsidy, the firm produces prototypes according to the market for which the innovations are most naturally suited. Introducing a subsidy causes the firm to rechannel prototypes from the private sector to the government. The dashed box represents the value (measured in investment dollars) of the innovations that are rechanneled toward the government—that is, the *influence* exerted by the GSI as a consequence of the rechanneling subsidy.

While the model does not explicitly distinguish the value of access to information that the GSI gives the government, the effect may be considered to be proportional to the amount of income the GSI generates for the government, and the government's relative interest in return on investment in fact reflects its interest in access to information. A more sophisticated extension of the model could consider incentives linked to information transfer. For example, portfolio companies might be more willing to share data if a GSI increased their sales. There is also the potential to create better incentives for the investment manager in the GSI to select

Figure 4.9
Equilibrium Outcomes as a Share of Total Budget

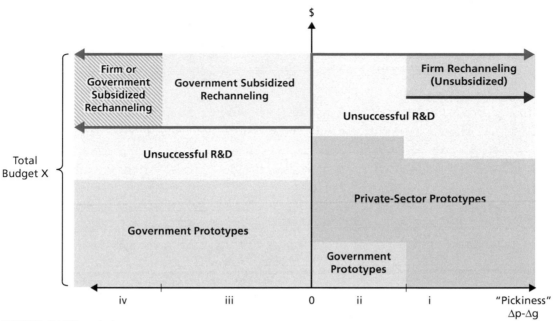

SOURCE: RAND analysis.

RAND RR176-4.9

Figure 4.10
Outcomes Without GSI Contract Funds to Subsidize Rechanneling

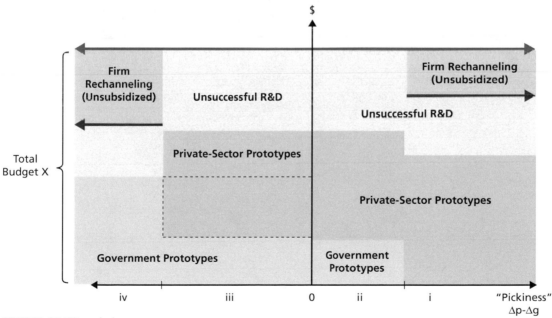

SOURCE: RAND analysis.

RAND RR176-4.10

a portfolio that (1) inherently favors government investment (γ) and that (2) identifies companies with non-picky private-sector markets that could be enticed to give the government more emphasis in their innovation (Δp).

Reviewing the insights gained from the economic model:

- When the private-sector customer is pickier than the government customer, no contract incentive should be applied to subsidize rechanneling toward government-specific prototype development.
- When the government sector is extremely sensitive to the specificity of the prototype, the firm will rechannel prototypes from the private sector to the government—independent of whether the GSI provides a contract incentive to subsidize rechanneling. If contract funds are not available, the cost of rechanneling will be paid from general funds.
- If the GSI exists at a point in the continuum where influence can be exerted on the firm (interval iii), the GSI will do so. The GSI's choice of rechanneling subsidy is increasing in the rate of innovation (λ), increasing in the cost of rechanneling (c), and decreasing in the share of innovation that is most naturally suited to government application (γ).

Furthermore, the model allows us to reflect on the relative performance of the GSIs examined in our case studies. In the case of OPT, all funds were designated for investment-supported R&D and no funds were set aside for contract incentives (d = 1). This strategy is appropriate when the private sector is pickier than the government (Δp > Δg) or when the government is *much* pickier than the private sector. The model suggests that a strategy of offering only contract incentives (d = 0), as was employed by RTVG, is rarely optimal because without funds allocated to investment-supported R&D, additional innovations cannot be generated. Finally, IQT pursued a hybrid innovation strategy that mixed both investment-supported R&D and contract incentives (0 < d < 1). This strategy is appropriate when the government is pickier than the private sector (Δg > Δp), independent of the magnitude of the difference (Δg - Δp).

CHAPTER FIVE

Observations from GSI Case Studies and Economic Modeling

Across the case study and economic framework tracks described in Chapters Three and Four respectively, RAND identified a number of general observations about the ways in which government strategic investment initiatives have been organized and operated, and how they managed incentives to innovate. These observations should have value for the design and management of future such initiatives.

Qualitative analysis of cases led to the following observations.

1. **In the three GSI cases examined, mission-oriented innovation was of equal or greater importance than generating financial return.** In each instance, significant effort was devoted to creating an organizational and legal framework that would provide direct benefit to accomplishing government mission objectives. Each GSI expended substantial effort to establish and maintain a good impedance match[1] between the private company providing the solution and its U.S. government investor/customer. In each case, there was an investment management organization to facilitate the match, turn expressions of customer need into investment proposals, conduct due diligence, and manage resources for both investment and development work programs.

2. **GSI participation in venture capital investments has provided government with additional information about technology-focused market sectors and companies.** The ability to participate directly in risk capital transactions has allowed the GSI investment managers to become part of the information sharing between entrepreneurs and private venture capitalists. The degree to which this information has translated into effective adoption of new technologies varies by case, and is not specifically evaluated in this research.

3. **GSI initiatives rely on the operational flexibility afforded by Other Transaction authority** (or "OT-like" authorities in the case of IQT) as a statutory foundation for both the contractual relationship with their sponsoring government agency investor and the contractual relationship they enter with private companies. OT authorities have allowed GSI investment managers great flexibility to combine investment with mission need-oriented prototype programs in ways that are specifically suited to the needs of individual companies on matters ranging from accounting practices and financial reporting to payments and intellectual property rights. Specific care is taken to struc-

---

[1] Impedance matching is the process of designing the input of a destination component to maximize power transfer from a source component. The term has specific technical meanings in electrical engineering, acoustics, optics, and mechanics, but can be applied to any situation where energy is transferred from a source to a destination.

ture the flow of government information rights consistent with the FAR to facilitate eventual scale adoption of prototypes.

4. **GSI initiatives rely in a significant way on a government-to-private sector "interface" function** that performs one or more of the following tasks: (1) providing contract administration, (2) identifying and "translating" investor mission-oriented needs into a form suitable for use by GSI investment managers, and (3) facilitating scale adoption of private company prototype solutions by mission-oriented government customers. These interface functions are performed by government employees, and serve to ensure that inherently government responsibilities dovetail appropriately with responsibilities discharged by GSI managers. GSI personnel do not have the organizational knowledge or breadth of expertise to assimilate all potential customer needs, and the interface organization typically includes employees of the government agency investor.

5. **The GSI's responsibility to government customers adds significant difficulty to the task of investment management.** The GSI must not only serve routine investment portfolio functions, such as identifying opportunities and negotiating and monitoring investments, but must also facilitate a good impedance match between government customers and the private companies in which the GSI invests, and do so in a way that does not confuse public and private responsibilities.

6. **GSI needs staff with private market capabilities to serve investment management functions.** It is difficult to overstate the importance of the quality and experience of the GSI investment management personnel. They serve a variety of functions, from devising investment theses to monitoring and harvesting investments. Even though staff monetary compensation for some GSI is lower than in private venture capital, the credibility necessary for staff to operate as peers in private investment transactions depends on them having skills equivalent to their private-sector counterparts.

Economic modeling analysis led to the following observations.

1. **Economic modeling can be used to help understand how alternative GSI incentive mechanisms (equity investment and contractual support) influence the technology development efforts of private firms.** This suggests a method for assessing alternative resource allocations that can be used both to design aspects of future GSI and to choose among incentive mechanisms depending on the degree to which government customers and the private-sector customers are sensitive to the specificity of an envisioned prototype. These sensitivities are likely to vary by technology and mission application area.

2. **The desired balance of GSI financial support between equity investment and contractual support depends on likelihood of sale in government and commercial markets.** GSI can exert influence on the quantity of innovations developed for government customers by setting aside funds for contractual support of development of government-specific prototypes, in cases where government customers are more sensitive to the specificity of the prototype than are private-sector customers. The flexibility inherent in OT authorities allows the GSI to balance its investment/contract offers to provide incentives for private companies to tailor innovation to address government mission objectives.

3. **The GSI initiatives in the case studies illustrate a range in the balance between equity investment and contractual support,** with OPT having most heavily emphasized the former, RTVG having most heavily emphasized the latter, and IQT having pursued a mixed strategy. Although this report does not examine the comparative effectiveness of these approaches, the economic analysis presents a framework within which to consider the suitable balance for future GSI initiatives. A more complete model would also consider incentives associated with information transfer, since these transfers are an important feature of GSI.

Economic Model Algebraic Details

The following pages contain detailed information regarding the economic model used to conduct the analysis in this report.

1   Model

There are two agents: the government venture capital (V) entity and the firm (F).

The exogenous variables are

- $X > 0$ is the total funds provided by the government venture capital

- $\lambda \in (0,1)$ is the probability with which each unit of R&D investment yields one unit of innovation

- $\gamma \in (0,1)$ is the share of innovation that naturally skews toward a government application

- $c_L = 0$ is the cost of developing one unit of innovation into a prototype, given that the prototype to be developed is the sort to which the innovation is naturally skewed. c_L is normalized to zero in order to focus on the cost of "re-channeling" innovation, as explained below.

- $c > 0$ is the cost of "re-channeling" innovation. More specifically, c is the cost of developing a prototype for the sector to which the innovation is *not* naturally skewed.

- $p_L \in [0,1)$ is the probability with which a prototype developed for the government can be sold to the private sector

- $p_H \in (p_L, 1]$ is the probability with which a prototype developed for the private sector can be sold to the private sector

- $g_L \in [0,1)$ is the probability with which a prototype developed for the private sector can be sold to the government

- $g_H \in (g_L, 1]$ is the probability with which a prototype developed for the government can be sold to the government

- $\theta \in [0,1]$ is the share of total sales that accrue to the government venture capital (i.e., the government venture capital's equity share)

- $\phi > 0$ is a scaling parameter that determines the extent to which the government venture capital benefits from government sales (independent of the benefit that accrues through the equity share)

The endogenous (choice) variables are

- $d \in [0, 1]$ is the share of total investment funds (X) the government venture capital designates for R&D investment

- $\beta \in [0, 1]$ is the share of total R&D investment funds (dX) the firm actually invests in R&D

- $\alpha_P \in [0, 1]$ is the share of total innovation ($\lambda \beta dX$) the firm develops into a prototype geared toward the private sector

- $\alpha_G \in [0, 1]$ is the share of total innovation ($\lambda \beta dX$) the firm develops into a prototype geared toward the government

The choice variable of the government venture capital is d. The choice variables of the firm are β, α_P, and α_G.

The timeline is as follows:

1. The government venture capital chooses d. The funds dX may be spent on either R&D investment or prototype development for either the private sector or the government. The funds $(1 - d)X$ must be spent on prototype development for the government.

2. The firm chooses β. The funds βdX are to be spent on R&D investment only. The funds $(1 - \beta)dX$ are to be spent on the development of prototypes for either the private sector or the government. The funds $(1 - d)X$ are to spent on the development of government prototypes only.

3. R&D investment occurs. The total quantity of innovation generated is $\lambda \beta dX$. The quantity of innovation that naturally skews toward a government application is $\gamma \lambda \beta dX$. The quantity of innovation that naturally skews toward a private sector application is $(1 - \gamma)\lambda \beta dX$.

4. The firm chooses α_P and α_G. The quantity of prototypes developed for the private sector is $\alpha_P \lambda \beta dX$. The quantity of prototypes developed for the government sector is $\alpha_G \lambda \beta dX$. Note that $\alpha_P + \alpha_G \leq 1$.

5. The firm incurs a re-channeling cost whenever it produces a prototype for a sector to which the underlying innovation is not naturally skewed. This cost depends on the the relative values of α_P, α_G, and γ.

   - Case 1: $\alpha_P \leq 1 - \gamma$ and $\alpha_G \leq \gamma$.

     There is no policy intervention. Prototypes are developed according to what they are naturally skewed to be. No re-channeling cost is incurred.

   - Case 2: $\alpha_P > 1 - \gamma$ and $\alpha_G \leq \gamma$.

     Some innovations are re-channeled from the government to the private sector. The re-channeling cost of producing these private sector prototypes is $[\alpha_P - (1 - \gamma)](\lambda \beta dX)c$.

   - Case 3: $\alpha_P \leq 1 - \gamma$ and $\alpha_G > \gamma$.

     Some innovations are re-channeled from the private sector to the government. The re-channeling cost of producing these government prototypes is $(\alpha_G - \gamma)(\lambda \beta dX)c$.

6. Sales occur. Total sales to the private sector are $(\alpha_P p_H + \alpha_G p_L)(\lambda \beta dX)$. Total sales to the government are $(\alpha_P g_L + \alpha_G g_H)(\lambda \beta dX)$.

7. The firm's utility is the sum of its sales times the equity share it retains: $U_F = (1-\theta)[\alpha_P(p_H + g_L) + \alpha_G(p_L + g_H)](\lambda \beta dX)$. The government venture capital's utility is the sum of sales times its equity share plus an additional benefit associated with having generated government sales: $U_V = \theta[\alpha_P(p_H + g_L) + \alpha_G(p_L + g_H)](\lambda \beta dX) + \phi(\alpha_P g_L + \alpha_G g_H)(\lambda \beta dX)$.

The following constraints must hold throughout:

- $\alpha_P + \alpha_G \leq 1$

- The re-channeling cost of producing private sector prototypes must be less than or equal to $(1 - \beta)dX$.

- The total re-channeling cost must be less than or equal to $(1-\beta)dX + (1-d)X$, or equivalently $(1 - \beta d)X$.

The following steps will be employed to solve the model:

1. Maximize U_F with respect to α_G. Use the solution to substitute for α_G in U_F.

2. Maximize U_F with respect to α_P. Use the solution to substitute for α_P in U_F.

3. Maximize U_F with respect to β. This yields β^*.

4. Implement the following substitutions:

   - Use β^* to substitute for β in α_P. This yields α_P^*.

   - Use α_P^* and β^* to substitute for α_P and β in α_G. This yields α_G^*.

   - Use α_G^*, α_P^*, and β^* to substitute for α_G, α_P, and β in U_F. This yields U_F^*.

   - Use α_G^*, α_P^*, and β^* to substitute for α_G, α_P, and β in U_V.

5. Maximize U_V with respect to d. This yields d^*.

6. Use d^* to substitute for d in U_V. This yields U_V^*.

7. Conduct comparative statics analyses.

The model will address the following questions:

1. What is the optimal distribution of funds (X) between R&D investment (d) and re-channeling innovations to government prototypes ($1 - d$)?

2. How does the share of funds designated for R&D investment (d) vary with

   - The probability of innovation (λ)?

   - The share of innovation that naturally skews toward government application (γ)?

   - The cost of re-channeling innovation (c)?

   - The relative probability of a sale to the government ($(g_H + g_L) - (p_H + p_L)$)? This may also be interpreted as the "closeness" between the government venture capital and the government customer(s).

- The relative probability of a sale of a government prototype $((g_H - g_L) - (p_H - p_L))$? This may also be interpreted as the extent to which the benefit of developing a sector-appropriate prototype varies across sectors.

- The importance of government sales (or government-specific innovation) to the government venture capital's mission (ϕ)?

3. Based on the answer to Question 2, what can be said about the conditions under which patents are preferred $(d = 1)$? Prizes are preferred $(d \to 1)$? A combination of patents and prizes is preferred?

4. What does the model tell us about the relative performance of the government venture capitals examined in our case studies?

- OnPoint: $d = 1$

- Rosettex: $d = 0$

- In-Q-Tel: $0 < d < 1$

5. What design (i.e., choice of d; choice of $g_H + g_L$ if possible) generates the optimal outcome for the government venture capital? What design generates the largest quantity of government sales (set $\theta = 0$)?

6. Under what conditions do firms decide to develop innovations for the government (α_G) rather than for the private sector (α_P)? How do these development shares vary with

- The share of funds designated for R&D investment (d)?

- The share of innovation that naturally skews toward government application (γ)?

- The cost of re-channeling innovation (c)?

- The relative probability of a sale to the government $((g_H + g_L) - (p_H + p_L))$? This may also be interpreted as the "closeness" between the government venture capital and the government customer(s).

- The relative probability of a sale of a government prototype $((g_H - g_L) - (p_H - p_L))$? This may also be interpreted as the extent to which the benefit of developing a sector-appropriate prototype varies across sectors.

Under what conditions does an increase in the share of funds designated for government re-channeling $(1 - d)$ induce an increase in the quantity of innovation developed for the government sector $(\alpha_G \lambda \beta dX)$?

2 Solving for α_G

In this section, we will maximize U_F with respect to α_G, holding all other variables (exogenous and endogenous) constant. Recall that the firm's utility is given by

$$U_F = (1 - \theta)[\alpha_P(p_H + g_L) + \alpha_G(p_L + g_H)](\lambda \beta dX),$$

which is strictly increasing in α_G. Hence, the firm would like to choose the largest α_G possible subject to the constraints. The analysis will be split into two cases. In Case 1, we assume $\alpha_P \leq 1 - \gamma$. In Case 2, we assume $\alpha_P > 1 - \gamma$.

2.1 Case 1: $\alpha_P \leq 1 - \gamma$

In this case, the constraints are as follows:

- $0 \leq \alpha_G \leq 1$

- $\alpha_P + \alpha_G \leq 1$

- If $\alpha_G > \gamma$, then $(\alpha_G - \gamma)(\lambda \beta dX)c \leq (1 - \beta d)X$

The approach will be motivated by the following intuition. If α_G can be set to $1 - \alpha_P$ without violating any of the constraints listed above, then α_G will be set in this fashion, which would imply that every innovation generated is developed into a prototype of some type. We will look for solutions of this sort first. If α_G cannot be set to $1 - \alpha_P$, it must be because the fourth and final constraint binds. This constraint is a budget constraint. Hence, solutions of this sort are such that the firm exhausts whatever budget remains for developing prototypes on the government prototypes.

Let us first derive the conditions under which $\alpha_G^* = 1 - \alpha_P$. Since $\alpha_P \leq 1 - \gamma$, setting α_G equal to $1 - \alpha_P$ implies that α_G will be greater than γ. Hence, the relevant budget constraint is

$$(\alpha_G - \gamma)(\lambda \beta dX)c \leq (1 - \beta d)X.$$

Substituting $1 - \alpha_P$ for α_G yields

$$\alpha_P \geq (1 - \gamma) - \frac{1 - \beta d}{(\lambda \beta d)c}.$$

The following lemma summarizes this result.

Lemma 1 *Suppose the following two conditions hold:*

(i) $\alpha_P \leq 1 - \gamma$ and

(ii) $\alpha_P \geq (1 - \gamma) - \dfrac{1 - \beta d}{(\lambda \beta d)c}.$

Then $\alpha_G^* = 1 - \alpha_P.$

Let us now consider the case in which

$$\alpha_P < (1 - \gamma) - \frac{1 - \beta d}{(\lambda \beta d)c}.$$

If $\alpha_G^* \leq \gamma$, there would be no budget constraint, and $\alpha_G^* = 1 - \alpha_P$. Hence, it must be the case that $\alpha_G^* > \gamma$. If the relevant budget constraint binds, we have

$$\alpha_G^* = \gamma + \frac{1 - \beta d}{(\lambda \beta d)c}.$$

The following lemma summarizes this result.

Lemma 2 *Suppose $\alpha_P < (1 - \gamma) - \dfrac{1 - \beta d}{(\lambda \beta d)c}$. Then $\alpha_G^* = \gamma + \dfrac{1 - \beta d}{(\lambda \beta d)c}.$*

This completes Case 1. In the following subsection, we consider the case in which $\alpha_P > 1 - \gamma$.

2.2   Case 2: $\alpha_P > 1 - \gamma$

In this case, the constraints are as follows:

- $0 \leq \alpha_G \leq 1$

- $\alpha_P + \alpha_G \leq 1$

- $[\alpha_P - (1 - \gamma)](\lambda \beta dX)c \leq (1 - \beta)dX$

- $[\alpha_P - (1 - \gamma)](\lambda \beta dX)c \leq (1 - \beta d)X$

Note that since $d \in [0, 1]$, the third constraint implies the fourth constraint. Also, note that in order for the third constraint to hold jointly with $\alpha_P > 1 - \gamma$, it must be the case that $\beta < 1$. Because the budget constraints represented by the third and fourth constraints apply to α_P but not α_G, it must be the case that $\alpha_G^* = 1 - \alpha_P$. This gives us our third and final lemma.

Lemma 3 *Suppose $\alpha_P > 1 - \gamma$. Then $\alpha_G^* = 1 - \alpha_P$.*

2.3   Solution Summary for α_G

Combining the results presented in Lemmas 1, 2, and 3, we obtain the solution for α_G and present it as Proposition 1.

Proposition 1 *Suppose either of the following two conditions holds:*

(i) $\alpha_P > 1 - \gamma$

(ii) $\alpha_P \leq 1 - \gamma$ and $\alpha_P \geq (1 - \gamma) - \dfrac{1 - \beta d}{(\lambda \beta d)c}$.

Then $\alpha_G^* = 1 - \alpha_P$. Suppose, instead, that the following condition holds:

(iii) $\alpha_P < (1 - \gamma) - \dfrac{1 - \beta d}{(\lambda \beta d)c}$.

Then $\alpha_G^* = \gamma + \dfrac{1 - \beta d}{(\lambda \beta d)c}$.

3   Solving for α_P

In this section, we will use the solution outlined in Proposition 1 to substitute for α_G in U_F and then maximize U_F with respect to α_P, holding all other variables (exogenous and endogenous) constant. Recall that the firm's utility is given by

$$U_F = A \left[\alpha_P(p_H + g_L) + \alpha_G(p_L + g_H) \right],$$

where $A \equiv (1 - \theta)(\lambda \beta d X)$. Solving for the α_P that maximizes U_F will depend on the relative magnitudes of the exogenous variables p_L, p_H, g_L, and g_H. For convenience, let $\Delta p \equiv p_H - p_L$ and $\Delta g \equiv g_H - g_L$. Δp represents the difference between the probability with which a private sector prototype can be sold to the private sector and the probability with which a government prototype can be sold to the private sector. Similarly, Δg represents the difference between the probability with which a government prototype can be sold to the government and the probability with which a private sector prototype can be sold to the government. The analysis will be split into three cases. In Case 1, we assume $\Delta p > \Delta g$, and in Case 2, we assume $\Delta g > \Delta p$. In Case 3, we assume $\Delta p = \Delta g$.

3.1   Case 1: $\Delta p > \Delta g$

Suppose $\alpha_G^* = 1 - \alpha_P$. Proposition 1 indicates this equality holds if either (i) $\alpha_P > 1 - \gamma$ or (ii) $\alpha_P \leq 1 - \gamma$ and $\alpha_P \geq (1 - \gamma) - (1 - \beta d)/(\lambda \beta dc)$. Substituting $1 - \alpha_P$ for α_G in U_F yields

$$U_F = A\left[(p_L + g_H) + \alpha_P(\Delta p - \Delta g)\right], \tag{3.1.1}$$

which is strictly increasing in α_P whenever $\Delta p > \Delta g$. Hence, the firm would like to choose the largest α_P possible subject to the constraints. If $\alpha_P > 1 - \gamma$ as in condition (i) of Proposition 1, the constraints are as follows:

- $\alpha_P > 1 - \gamma$

- $0 \leq \alpha_P \leq 1$

- $[\alpha_P - (1 - \gamma)](\lambda \beta dX)c \leq (1 - \beta)dX$

Maximizing α_P subject to these constraints yields $\alpha_P^* = \min\{(1 - \gamma) + (1 - \beta)/(\lambda \beta c), 1\}$. Substituting for α_P in equation (**??**) yields

$$U_F^* = \begin{cases} A\left[\left((1 - \gamma) + \dfrac{1 - \beta}{\lambda \beta c}\right)(p_H + g_L) & \text{if } \gamma > \dfrac{1 - \beta}{\lambda \beta c} \\ \qquad + \left(\gamma - \dfrac{1 - \beta}{\lambda \beta c}\right)(p_L + g_H)\right] \\[2mm] A\,(p_H + g_L) & \text{if } \gamma \leq \dfrac{1 - \beta}{\lambda \beta c} \end{cases} \tag{3.1.2}$$

If instead $\alpha_P \leq 1 - \gamma$ and $\alpha_P \geq (1 - \gamma) - (1 - \beta d)/(\lambda \beta dc)$ as in condition (ii) of Proposition 1, the constraints are as follows:

- $\alpha_P \leq 1 - \gamma$

- $\alpha_P \geq (1 - \gamma) - (1 - \beta d)/(\lambda \beta dc)$

- $0 \leq \alpha_P \leq 1$

Maximizing α_P subject to these constraints yields $\alpha_P^* = 1 - \gamma$. Substituting for α_P in equation (??) yields

$$U_F^* = A\left[(1-\gamma)(p_H + g_L) + \gamma(p_L + g_H)\right]. \tag{3.1.3}$$

Now suppose $\alpha_G^* = \gamma + (1 - \beta d)/(\lambda \beta dc)$. Proposition 1 indicates this equality holds if $\alpha_P < (1-\gamma) - (1 - \beta d)/(\lambda \beta dc)$. Substituting for α_G in U_F yields

$$U_F = A\left[\alpha_P(p_H + g_L) + \left(\gamma + \frac{1 - \beta d}{(\lambda \beta d)c}\right)(p_L + g_H)\right], \tag{3.1.4}$$

which is strictly increasing in α_P. Hence, the firm would like to choose the largest α_P possible subject to the constraints. If $\alpha_P < (1-\gamma) - (1 - \beta d)/(\lambda \beta dc)$ as in condition (iii) of Proposition 1, the constraints are as follows:

- $\alpha_P < (1-\gamma) - \dfrac{1 - \beta d}{(\lambda \beta d)c}$

- $0 \leq \alpha_P \leq 1$

Note that in order for the two constraints to hold jointly, parameter values must be such that $1 - \gamma > (1 - \beta d)/(\lambda \beta dc)$. Maximizing α_P subject to the two constraints yields $\alpha_P^* = (1-\gamma) - (1 - \beta d)/(\lambda \beta dc) - \epsilon$, where $\epsilon > 0$ is arbitrarily small. Substituting for α_P in equation (??) yields

$$U_F^* = A\left[\left((1-\gamma)\frac{1 - \beta d}{(\lambda \beta d)c} - \epsilon\right)(p_H + g_L) + \left(\gamma + \frac{1 - \beta d}{(\lambda \beta d)c}\right)(p_L + g_H)\right]. \tag{3.1.5}$$

We will now compare equations (??), (??), and (??) to identify the α_P that maximizes the firm's utility. Since $\Delta p > \Delta g$, it must be the case that $p_H + g_L > p_L + g_H$. It follows that the utility level expressed in equation (??) is strictly larger than the utility level expressed in equation (??). Hence, we can assert that in equilibrium $\alpha_G^* = 1 - \alpha_P^*$. Moreover, the utility levels expressed in equation (??) are strictly greater than the utility level expressed in equation (??). Hence, in equilibrium $\alpha_P^* = \min\{(1-\gamma) + (1 - \beta)/(\lambda \beta c), 1\}$. The following lemma summarizes our results for Case 1.

Lemma 4 *Suppose $\Delta p > \Delta g$. Then $\alpha_P^* = \min\left\{(1-\gamma) + \dfrac{1 - \beta}{\lambda \beta c}, 1\right\}$ and $\alpha_G^* = 1 - \alpha_P^*$.*

3.2   Case 2: $\Delta g > \Delta p$

Suppose $\alpha_G^* = 1 - \alpha_P$. Proposition 1 indicates this equality holds if either (i) $\alpha_P > 1 - \gamma$ or (ii) $\alpha_P \leq 1 - \gamma$ and $\alpha_P \geq (1 - \gamma) - (1 - \beta d)/(\lambda\beta dc)$. Substituting $1 - \alpha_P$ for α_G in U_F yields

$$U_F = A\left[(p_L + g_H) + \alpha_P(\Delta p - \Delta g)\right], \tag{3.2.1}$$

which is strictly decreasing in α_P whenever $\Delta g > \Delta p$. Hence, the firm would like to choose the smallest α_P possible subject to the constraints. If $\alpha_P > 1 - \gamma$ as in condition (i) of Proposition 1, the constraints are as follows:

- $\alpha_P > 1 - \gamma$

- $0 \leq \alpha_P \leq 1$

- $[\alpha_P - (1 - \gamma)](\lambda\beta dX)c \leq (1 - \beta)dX$

Minimizing α_P subject to these constraints yields $\alpha_P^* = (1 - \gamma) + \epsilon$, where $\epsilon > 0$ is arbitrarily small. Substituting for α_P in equation (??) yields

$$U_F^* = A\left[(1 - \gamma + \epsilon)(p_H + g_L) + (\gamma - \epsilon)(p_L + g_H)\right] \tag{3.2.2}$$

If instead $\alpha_P \leq 1 - \gamma$ and $\alpha_P \geq (1 - \gamma) - (1 - \beta d)/(\lambda\beta dc)$ as in condition (ii) of Proposition 1, the constraints are as follows:

- $\alpha_P \leq 1 - \gamma$

- $\alpha_P \geq (1 - \gamma) - (1 - \beta d)/(\lambda\beta dc)$

- $0 \leq \alpha_P \leq 1$

Minimizing α_P subject to these constraints yields $\alpha_P^* = \max\{0, (1 - \gamma) - (1 - \beta d)/(\lambda\beta dc)\}$. Substituting for α_P in equation (??) yields

$$U_F^* = \begin{cases} A\left[\left((1 - \gamma) - \dfrac{1 - \beta d}{(\lambda\beta d)c}\right)(p_H + g_L) \right. & \text{if } 1 - \gamma > \dfrac{1 - \beta d}{(\lambda\beta d)c} \\ \left. \qquad + \left(\gamma + \dfrac{1 - \beta d}{(\lambda\beta d)c}\right)(p_L + g_H)\right] & \\ \\ A\,(p_L + g_H) & \text{if } 1 - \gamma \leq \dfrac{1 - \beta d}{(\lambda\beta d)c} \end{cases} \tag{3.2.3}$$

Now suppose $\alpha_G^* = \gamma + (1 - \beta d)/(\lambda \beta dc)$. Proposition 1 indicates this equality holds if $\alpha_P < (1 - \gamma) - (1 - \beta d)/(\lambda \beta dc)$. Substituting for α_G in U_F yields

$$U_F = A\left[\alpha_P(p_H + g_L) + \left(\gamma + \frac{1 - \beta d}{(\lambda \beta d)c}\right)(p_L + g_H)\right], \tag{3.2.4}$$

which is strictly increasing in α_P. Hence, the firm would like to choose the largest α_P possible subject to the constraints. If $\alpha_P < (1 - \gamma) - (1 - \beta d)/(\lambda \beta dc)$ as in condition (iii) of Proposition 1, the constraints are as follows:

- $\alpha_P < (1 - \gamma) - \dfrac{1 - \beta d}{(\lambda \beta d)c}$

- $0 \le \alpha_P \le 1$

Note that in order for the two constraints to hold jointly, parameter values must be such that $1 - \gamma > (1 - \beta d)/(\lambda \beta dc)$. Maximizing α_P subject to the two constraints yields $\alpha_P^* = (1 - \gamma) - (1 - \beta d)/(\lambda \beta dc) - \epsilon$, where $\epsilon > 0$ is arbitrarily small. Substituting for α_P in equation (??) yields

$$U_F^* = A\left[\left((1 - \gamma) - \frac{1 - \beta d}{(\lambda \beta d)c} - \epsilon\right)(p_H + g_L) + \left(\gamma + \frac{1 - \beta d}{(\lambda \beta d)c}\right)(p_L + g_H)\right]. \tag{3.2.5}$$

We will now compare equations (??), (??), and (??) to identify the α_P that maximizes the firm's utility. Since $\Delta g > \Delta p$, it must be the case that $p_L + g_H > p_H + g_L$. It follows that the utility level expressed in equation (??) is strictly larger than the utility level expressed in equation (??). Hence, we can assert that in equilibrium $\alpha_G^* = 1 - \alpha_P^*$. Moreover, the utility levels expressed in equation (??) are strictly greater than the utility level expressed in equation (??). Hence, in equilibrium $\alpha_P^* = \max\{0, (1 - \gamma) - (1 - \beta d)/(\lambda \beta dc)\}$. The following lemma summarizes our results for Case 2.

Lemma 5 *Suppose $\Delta g > \Delta p$. Then $\alpha_P^* = \max\left\{0, (1 - \gamma) - \dfrac{1 - \beta d}{(\lambda \beta d)c}\right\}$ and $\alpha_G^* = 1 - \alpha_P^*$.*

3.3 Case 3: $\Delta p = \Delta g$

If $\Delta p = \Delta g$, then $p_H + g_L = p_L + g_H$. Let $P \equiv p_H + g_L$. The firm's utility can then be written as

$$\begin{aligned} U_F &= A[\alpha_P(p_H + g_L) + \alpha_G(p_L + g_H)] \\ &= AP(\alpha_P + \alpha_G) \end{aligned} \tag{3.3.1}$$

Suppose $\alpha_G^* = 1 - \alpha_P$. Proposition 1 indicates this equality holds if either (i) $\alpha_P > 1 - \gamma$ or (ii) $\alpha_P \le 1 - \gamma$ and $\alpha_P \ge (1 - \gamma) - (1 - \beta d)/(\lambda \beta dc)$. Substituting $1 - \alpha_P$ for α_G in equation (??)

yields $U_F = AP$, which does not vary with α_P. Hence, the firm is indifferent among any α_P that satisfies either

- $\alpha_P > 1 - \gamma$,

- $0 \leq \alpha_P \leq 1$, and

- $[\alpha_P - (1-\gamma)](\lambda\beta dX)c \leq (1-\beta)dX$

or

- $\alpha_P \leq 1 - \gamma$,

- $\alpha_P \geq (1-\gamma) - (1-\beta d)/(\lambda\beta dc)$, and

- $0 \leq \alpha_P \leq 1$.

That is, the firm is indifferent among any $\alpha_P \in [\underline{\alpha}, \overline{\alpha}]$, where $\underline{\alpha} \equiv \max\{0, (1-\gamma) - (1-\beta d)/(\lambda\beta dc)\}$ and $\overline{\alpha} \equiv \min\{(1-\gamma) + (1-\beta)/(\lambda\beta c), 1\}$. Any such α_P yields $U_F^* = AP$.

Now suppose $\alpha_G^* = \gamma + (1-\beta d)/(\lambda\beta dc)$. Proposition 1 indicates this equality holds if $\alpha_P < (1-\gamma) - (1-\beta d)/(\lambda\beta dc)$. Substituting for α_G in equation (??) yields

$$U_F = AP\left[\alpha_P + \left(\gamma + \frac{1-\beta d}{(\lambda\beta d)c}\right)\right], \tag{3.3.2}$$

which is strictly increasing in α_P. Hence, the firm would like to choose the largest α_P possible subject to the constraints. If $\alpha_P < (1-\gamma) - (1-\beta d)/(\lambda\beta dc)$ as in condition (iii) of Proposition 1, the constraints are as follows:

- $\alpha_P < (1-\gamma) - \dfrac{1-\beta d}{(\lambda\beta d)c}$

- $0 \leq \alpha_P \leq 1$

Note that in order for the two constraints to hold jointly, parameter values must be such that $1 - \gamma > (1-\beta d)/(\lambda\beta dc)$. Maximizing α_P subject to the two constraints yields $\alpha_P^* = (1-\gamma) - (1-\beta d)/(\lambda\beta dc) - \epsilon$, where $\epsilon > 0$ is arbitrarily small. Substituting for α_P in equation (??) yields $U_F^* = AP(1-\epsilon)$.

We will now compare $U_F^* = AP$ and $U_F^* = AP(1-\epsilon)$ to identify the α_P that maximizes the firm's utility. The former is clearly greater. Hence, in equilibrium α_P^* may take any value in $[\underline{\alpha}, \overline{\alpha}]$. The following lemma summarizes our results for Case 3.

Lemma 6 *Suppose* $\Delta p = \Delta g$. *Then* $\alpha_P^* \in [\underline{\alpha}, \overline{\alpha}]$, *where* $\underline{\alpha} = \max \left\{ 0, (1-\gamma) - \dfrac{1-\beta d}{(\lambda \beta d)c} \right\}$ *and* $\overline{\alpha} = \min \left\{ (1-\gamma) + \dfrac{1-\beta}{\lambda \beta c}, 1 \right\}$. *Moreover,* $\alpha_G^* = 1 - \alpha_P^*$.

3.4   Solution Summary for α_P

Combining the results in Lemmas 4, 5, and 6, we obtain the solution for α_P and present it as Proposition 2.

Proposition 2 *Let* α_P^* *and* α_G^* *represent the equilibrium values of* α_P *and* α_G *respectively.*

(i) If $\Delta p > \Delta g$, *then* $\alpha_P^* = \min \left\{ (1-\gamma) + \dfrac{1-\beta}{\lambda \beta c}, 1 \right\}$.

(ii) If $\Delta g > \Delta p$, *then* $\alpha_P^* = \max \left\{ 0, (1-\gamma) - \dfrac{1-\beta d}{(\lambda \beta d)c} \right\}$.

(iii) If $\Delta p = \Delta g$, *then* $\alpha_P^* \in [\underline{\alpha}, \overline{\alpha}]$,

where $\underline{\alpha} = \max \left\{ 0, (1-\gamma) - \dfrac{1-\beta d}{(\lambda \beta d)c} \right\}$ *and* $\alpha = \min \left\{ (1-\gamma) + \dfrac{1-\beta}{\lambda \beta c}, 1 \right\}$.

In all three cases, $\alpha_G^* = 1 - \alpha_P$.

4   Solving for β

In this section, we will use the solution outlined in Proposition 2 to substitute for α_P and α_G in U_F and then maximize U_F with respect to β, holding all other variables (exogenous and endogenous) constant. Following Proposition 2, the analysis will be split into three cases. In Case 1, we assume $\Delta p > \Delta g$; in Case 2, we assume $\Delta g > \Delta p$; and in Case 3, we assume $\Delta p = \Delta g$.

4.1   Case 1: $\Delta p > \Delta g$

Proposition 2 indicates that if $\Delta p > \Delta g$, then $\alpha_P^* = \min\{(1-\gamma) + (1-\beta)/(\lambda \beta c), 1\}$ and $\alpha_G^* = 1 - \alpha_P$. Substituting α_P^* and α_G^* for α_P and α_G in U_F yields

$$U_F = \begin{cases} A\,(p_H + g_L) & \text{if } \gamma \leq \dfrac{1-\beta}{\lambda \beta c} \\[2em] A\left[\left((1-\gamma) + \dfrac{1-\beta}{\lambda \beta c} \right)(p_H + g_L) \right. & \\[1em] \left. + \left(\gamma - \dfrac{1-\beta}{\lambda \beta c} \right)(p_L + g_H) \right] & \text{if } \gamma > \dfrac{1-\beta}{\lambda \beta c} \end{cases} \qquad (4.1.1)$$

where $A = (1 - \theta)(\lambda \beta dX)$.

Suppose $\gamma \leq (1 - \beta)/(\lambda \beta c)$. Then the partial derivative of U_F with respect to β is

$$\frac{\partial U_F}{\partial \beta} = (1 - \theta)(\lambda dX)(p_H + g_L),$$

which is strictly positive. Hence, the firm would like to choose the largest β possible subject to the constraints:

- $0 \leq \beta \leq 1$

- $\gamma \leq (1 - \beta)/(\lambda \beta c)$

- $[\alpha_P - (1 - \gamma)](\lambda \beta dX)c \leq (1 - \beta)dX$

- $[\alpha_P - (1 - \gamma)](\lambda \beta dX)c \leq (1 - \beta d)X$

Note that since $d \in [0, 1]$, the third constraint implies the fourth constraint. Moreover, if the second constraint holds, then $\alpha_P^* = 1$, and the third constraint reduces to the second constraint. Hence, the set of four constraints listed above reduces to two: $\gamma \leq (1 - \beta)/(\lambda \beta c)$ and $0 \leq \beta \leq 1$. Maximizing β subject to these constraints yields $\beta^* = 1/(1 + \gamma \lambda c)$. Substituting for β in U_F yields

$$U_F^* = (1 - \theta)(\lambda dX)\left(\frac{1}{1 + \gamma \lambda c}\right)(p_H + g_L). \tag{4.1.2}$$

Suppose $\gamma > (1 - \beta)/(\lambda \beta c)$. Then the partial derivative of U_F with respect to β is

$$\frac{\partial U_F}{\partial \beta} = (1 - \theta)(\lambda dX)\left[(p_H + g_L) - \left(\frac{1 + \gamma \lambda c}{\lambda c}\right)(\Delta p - \Delta g)\right].$$

If $\Delta p - \Delta g < [(\lambda c)/(1 + \gamma \lambda c)](p_H + g_L)$, then the partial derivative is strictly positive, and the firm would like to choose the largest β possible subject to the constraints:

- $0 \leq \beta \leq 1$

- $\gamma > (1 - \beta)/(\lambda \beta c)$

- $[\alpha_P - (1 - \gamma)](\lambda \beta dX)c \leq (1 - \beta)dX$

If the second constraint holds, then $\alpha_P^* = (1 - \gamma) + (1 - \beta)/(\lambda \beta c)$, which ensures that the third constraint is satisfied. Hence, the set of three constraints listed above reduces to two: $\gamma > (1 -$

$\beta)/(\lambda\beta c)$ and $0 \leq \beta \leq 1$. Maximizing β subject to these constraints yields $\beta^* = 1$. Substituting for β in U_F yields

$$U_F^* = (1 - \theta)(\lambda dX)\left[(1 - \gamma)(p_H + g_L) + \gamma(p_L + g_H)\right]. \tag{4.1.3}$$

If $\Delta p - \Delta g > [(\lambda c)/(1 + \gamma\lambda c)](p_H + g_L)$, then the partial derivative is strictly negative, and the firm would like to choose the smallest β possible subject to the constraints: $\gamma > (1 - \beta)/(\lambda\beta c)$ and $0 \leq \beta \leq 1$. This minimization yields $\beta^* = 1/(1 + \gamma\lambda c) + \epsilon$, where $\epsilon > 0$ is arbitrarily small. Substituting for β in U_F yields

$$U_F^* = (1 - \theta)(\lambda dX)\left(\frac{1}{1 + \gamma\lambda c}\right)(p_H + g_L) - \delta, \tag{4.1.4}$$

where $\delta > 0$ is arbitrarily small. More specifically, $\delta = (1 - \theta)(\lambda dX)([(1 + \gamma\lambda c)/(\lambda c)](\Delta p - \Delta g) - (p_H + g_L))\epsilon$. If $\Delta p - \Delta g = [(\lambda c)/(1 + \gamma\lambda c)](p_H + g_L)$, then the partial derivative is zero, and the firm is indifferent among any β that satisfies the constraints: $\gamma > (1 - \beta)/(\lambda\beta c)$ and $0 \leq \beta \leq 1$. Hence, $\beta^* \in (1/(1 + \gamma\lambda c), 1]$. Substituting for β in U_F yields

$$U_F^* = (1 - \theta)\left(\frac{dX}{c}\right)(\Delta p - \Delta g). \tag{4.1.5}$$

We will now compare equation (**??**) with each of the three subcases: equations (**??**), (**??**), and (**??**). If $\Delta p - \Delta g < [(\lambda c)/(1 + \gamma\lambda c)](p_H + g_L)$, then the appropriate comparison is between equations (**??**) and (**??**). Since the utility level expressed in equation (**??**) is strictly greater than the utility level expressed in equation (**??**), we can assert that in equilibrium $\beta^* = 1$. If $\Delta p - \Delta g > [(\lambda c)/(1 + \gamma\lambda c)](p_H + g_L)$, then the appropriate comparison is between equations (**??**) and (**??**). Since the utility level expressed in equation (**??**) is strictly greater than the utility level expressed in equation (**??**), we can assert that in equilibrium $\beta^* = 1/(1 + \gamma\lambda c)$. If $\Delta p - \Delta g = [(\lambda c)/(1 + \gamma\lambda c)](p_H + g_L)$, then the appropriate comparison is between equations (**??**) and (**??**). Since the utility levels expressed in the two equations are equal, we can assert that in equilibrium β^* may take any value in $[1/(1 + \gamma\lambda c), 1]$.

The following lemma summarizes our results for Case 1.

Lemma 7 *Suppose $\Delta p > \Delta g$.*

(i) If $\Delta p - \Delta g < \left(\dfrac{\lambda c}{1 + \gamma\lambda c}\right)(p_H + g_L)$, then $\beta^* = 1$.

(ii) If $\Delta p - \Delta g > \left(\dfrac{\lambda c}{1 + \gamma\lambda c}\right)(p_H + g_L)$, then $\beta^* = \dfrac{1}{1 + \gamma\lambda c}$.

(iii) If $\Delta p - \Delta g = \left(\dfrac{\lambda c}{1 + \gamma \lambda c}\right)(p_H + g_L),$ *then* $\beta^* \in \left[\dfrac{1}{1 + \gamma \lambda c}, 1\right].$

In all three cases, $\alpha_P^* = \min\left\{(1 - \gamma) + \dfrac{1 - \beta}{\lambda \beta c}, 1\right\}$ *and* $\alpha_G^* = 1 - \alpha_P.$

4.2   Case 2: $\Delta g > \Delta p$

Proposition 2 indicates that if $\Delta g > \Delta p$, then $\alpha_P^* = \max\{0, (1 - \gamma) - (1 - \beta d)/(\lambda \beta dc)\}$ and $\alpha_G^* = 1 - \alpha_P$. Substituting α_P^* and α_G^* for α_P and α_G in U_F yields

$$U_F = \begin{cases} A\left(p_L + g_H\right) & \text{if } 1 - \gamma \leq \dfrac{1 - \beta d}{(\lambda \beta d)c} \\[2em] A\left[\left((1 - \gamma) - \dfrac{1 - \beta d}{(\lambda \beta d)c}\right)(p_H + g_L)\right. \\[1em] \left. + \left(\gamma + \dfrac{1 - \beta d}{(\lambda \beta d)c}\right)(p_L + g_H)\right] & \text{if } 1 - \gamma > \dfrac{1 - \beta d}{(\lambda \beta d)c} \end{cases} \tag{4.2.1}$$

where $A = (1 - \theta)(\lambda \beta dX)$.

Suppose $1 - \gamma \leq (1 - \beta d)/(\lambda \beta dc)$. Then the partial derivative of U_F with respect to β is

$$\frac{\partial U_F}{\partial \beta} = (1 - \theta)(\lambda dX)(p_L + g_H),$$

which is strictly positive. Hence, the firm would like to choose the largest β possible subject to the constraints:

- $0 \leq \beta \leq 1$

- $1 - \gamma \leq (1 - \beta d)/(\lambda \beta dc)$

- $(\alpha_G - \gamma)(\lambda \beta dX)c \leq (1 - \beta d)X$

If the second constraint holds, then $\alpha_P^* = 0$ and $\alpha_P^* = 1$, which implies that the third constraint collapses to the second constraint. Hence, the set of three constraints listed above reduces to two: $1 - \gamma \leq (1 - \beta d)/(\lambda \beta dc)$ and $0 \leq \beta \leq 1$. Maximizing β subject to these constraints yields $\beta^* = \min\{1/(d[1 + (1 - \gamma)\lambda c]), 1\}$. Substituting for β in U_F yields

$$U_F^* = \begin{cases} (1 - \theta)(\lambda dX)(p_L + g_H) & \text{if } d[1 + (1 - \gamma)\lambda c] \leq 1 \\[2em] (1 - \theta)\left(\dfrac{\lambda X}{1 + (1 - \gamma)\lambda c}\right)(p_L + g_H) & \text{if } d[1 + (1 - \gamma)\lambda c] > 1 \end{cases} \tag{4.2.2}$$

Suppose $1 - \gamma > (1 - \beta d)/(\lambda \beta dc)$. In order for this inequality to hold, it must be the case that $d\left[1 + (1 - \gamma)\lambda c\right] > 1$. The partial derivative of U_F with respect to β is

$$\frac{\partial U_F}{\partial \beta} = (1 - \theta)(\lambda dX)\left[(p_H + g_L) - \left(\frac{1 - \gamma \lambda c}{\lambda c}\right)(\Delta g - \Delta p)\right].$$

If $p_H + g_L > [(1 - \gamma \lambda c)/(\lambda c)](\Delta g - \Delta p)$, then the partial derivative is strictly positive, and the firm would like to choose the largest β possible subject to the constraints:

- $0 \leq \beta \leq 1$

- $1 - \gamma > (1 - \beta d)/(\lambda \beta dc)$

- $(\alpha_G - \gamma)(\lambda \beta dX)c \leq (1 - \beta d)X$

If the second constraint holds, then $\alpha_P^* = (1 - \gamma) - (1 - \beta d)/(\lambda \beta dc)$ and $\alpha_G^* = \gamma + (1 - \beta d)/(\lambda \beta dc)$, which ensures that the third constraint is satisfied. Hence, the set of three constraints listed above reduces to two: $1 - \gamma > (1 - \beta d)/(\lambda \beta dc)$ and $0 \leq \beta \leq 1$. Maximizing β subject to these constraints yields $\beta^* = 1$. Substituting for β in U_F yields

$$U_F^* = (1 - \theta)(\lambda dX)\left[\left((1 - \gamma) - \frac{1 - d}{\lambda dc}\right)(p_H + g_L) + \left(\gamma + \frac{1 - d}{\lambda dc}\right)(p_L + g_H)\right] \qquad (4.2.3)$$

If $p_H + g_L < [(1 - \gamma \lambda c)/(\lambda c)](\Delta g - \Delta p)$, then the partial derivative is strictly negative, and the firm would like to choose the smallest β possible subject to the constraints: $1 - \gamma > (1 - \beta d)/(\lambda \beta dc)$ and $0 \leq \beta \leq 1$. This minimization yields $\beta^* = 1/(d[1 + (1 - \gamma)\lambda c]) + \epsilon$, where $\epsilon > 0$ is arbitrarily small. Substituting for β in U_F yields

$$U_F^* = (1 - \theta)\left(\frac{\lambda X}{1 + (1 - \gamma)\lambda c}\right)(p_L + g_H) - \delta, \qquad (4.2.4)$$

where $\delta > 0$ is arbitrarily small. More specifically, $\delta = (1 - \theta)(\lambda X)([(1 - \gamma \lambda c)/(\lambda c)](\Delta g - \Delta p) - (p_H + g_L))\epsilon$. If $p_H + g_L = [(1 - \gamma \lambda c)/(\lambda c)](\Delta g - \Delta p)$, then the partial derivative is zero, and the firm is indifferent among any β that satisfies the constraints: $1 - \gamma > (1 - \beta d)/(\lambda \beta dc)$ and $0 \leq \beta \leq 1$. Hence, $\beta^* \in (1/(d[1 + (1 - \gamma)\lambda c]), 1]$. Substituting for β in U_F yields

$$U_F^* = (1 - \theta)\left(\frac{X}{c}\right)(\Delta g - \Delta p). \qquad (4.2.5)$$

We will now compare equation (??) with each of the three subcases: equations (??), (??), and (??). Suppose $d[1 + (1 - \gamma)\lambda c] \leq 1$. Since none of the three subcases are viable, we can assert that in equilibrium $\beta^* = 1$. Suppose instead that $d[1+(1-\gamma)\lambda c] > 1$. If $p_H+g_L > [(1-\gamma\lambda c)/(\lambda c)](\Delta g-\Delta p)$, then the appropriate comparison is between equations (??) and (??). The utility level expressed in equation (??) is

$$(1 - \theta) \left(\frac{\lambda X}{1 + (1 - \gamma)\lambda c} \right) (p_L + g_H),$$

which is the same utility level one obtains when evaluating U_F at $\beta = 1/(d[1 + (1 - \gamma)\lambda c])$. Since the second branch of equation (??) is strictly increasing in β, the utility level expressed in equation (??) is strictly greater than the utility level expressed in equation (??). Hence, we can assert that in equilibrium $\beta^* = 1$. If $p_H + g_L < [(1 - \gamma\lambda c)/(\lambda c)](\Delta g - \Delta p)$, then the appropriate comparison is between equations (??) and (??). Since the utility level expressed in equation (??) is strictly greater than the utility level expressed in equation (??), we can assert that in equilibrium $\beta^* = 1/(d[1+(1-\gamma)\lambda c])$. If $p_H+g_L = [(1-\gamma\lambda c)/(\lambda c)](\Delta g-\Delta p)$, then the appropriate comparison is between equations (??) and (??). Since the utility levels expressed in the two equations are equal, we can assert that in equilibrium β^* may take any value in $[1/(d[1 + (1 - \gamma)\lambda c]), 1]$.

The following lemma summarizes our results for Case 2.

Lemma 8 *Suppose $\Delta g > \Delta p$.*

(i) If $d\,[1 + (1 - \gamma)\lambda c] \le 1$, then $\beta^* = 1$.

(ii) If $d\,[1 + (1 - \gamma)\lambda c] > 1$ and $p_H + g_L > \left(\dfrac{1 - \gamma\lambda c}{\lambda c} \right)(\Delta g - \Delta p)$, then $\beta^* = 1$.

(iii) If $d\,[1 + (1 - \gamma)\lambda c] > 1$ and $p_H + g_L < \left(\dfrac{1 - \gamma\lambda c}{\lambda c} \right)(\Delta g - \Delta p)$, then $\beta^* = \dfrac{1}{d\,[1 + (1 - \gamma)\lambda c]}$.

(iv) If $d\,[1 + (1 - \gamma)\lambda c] > 1$ and $p_H+g_L = \left(\dfrac{1 - \gamma\lambda c}{\lambda c} \right)(\Delta g-\Delta p)$, then $\beta^* \in \left[\dfrac{1}{d\,[1 + (1 - \gamma)\lambda c]}, 1 \right]$.

In all four cases, $\alpha_P^* = \max\left\{ 0, (1 - \gamma) - \dfrac{1 - \beta d}{(\lambda\beta d)c} \right\}$ and $\alpha_G^* = 1 - \alpha_P$.

4.3   Case 3: $\Delta p = \Delta g$

Proposition 2 indicates that if $\Delta p = \Delta g$, then $\alpha_P^* \in [\underline{\alpha}, \overline{\alpha}]$ and $\alpha_G^* = 1 - \alpha_P$, where $\underline{\alpha} = \max\{0, (1 - \gamma) - (1 - \beta d)/(\lambda\beta dc)\}$ and $\overline{\alpha} = \min\{(1 - \gamma) + (1 - \beta)/(\lambda\beta c), 1\}$. Substituting α_P^* and α_G^* for α_P and α_G in U_F yields

$$U_F = A(p_H + g_L), \tag{4.3.1}$$

where $A = (1-\theta)(\lambda\beta dX)$. The partial derivative of U_F with respect to β is

$$\frac{\partial U_F}{\partial \beta} = (1-\theta)(\lambda dX)(p_H + g_L),$$

which is strictly positive. Hence, the firm would like to choose the largest β possible subject to the constraints:

- $0 \le \beta \le 1$

- If $\alpha_P > 1 - \gamma$, then $[\alpha_P - (1-\gamma)](\lambda\beta dX)c \le (1-\beta)dX$

- If $\alpha_P \le 1 - \gamma$, then $[(1-\gamma) - \alpha_P](\lambda\beta dX)c \le (1-\beta d)X$

We will now verify that $\beta^* = 1$ is feasible given the constraints listed. If $\beta^* = 1$, then $\alpha_P^* \in [\underline{\alpha}^*, 1-\gamma]$, where $\underline{\alpha}^* = \max\{0, (1-\gamma)-(1-d)/(\lambda dc)\}$. The first constraint listed above is satisfied trivially, and the second constraint does not apply. The third constraint reduces to $\alpha_P^* \ge (1-\gamma) - (1-d)/(\lambda dc)$, which is satisfied by any α_P^* drawn from $[\underline{\alpha}^*, 1-\gamma]$. The following lemma summarizes our results for Case 3.

Lemma 9 *Suppose* $\Delta p = \Delta g$. *Then* $\beta^* = 1$, $\alpha_P^* \in \left[\max\left\{0, (1-\gamma) - \dfrac{1-d}{\lambda dc}\right\}, 1-\gamma\right]$, *and* $\alpha_G^* = 1 - \alpha_P$.

4.4 Solution Summary for β

Combining the results in Lemmas 7, 8, and 9, we obtain the solution for β and present it as Proposition 3.

Proposition 3 *Let* β^*, α_P^*, *and* α_G^* *represent the equilibrium values of* β, α_P, *and* α_G *respectively.*

(i) If $\Delta p - \Delta g > \left(\dfrac{\lambda c}{1+\gamma\lambda c}\right)(p_H + g_L)$, *then* $\beta^* = \dfrac{1}{1+\gamma\lambda c}$.

(ii) If $\Delta p - \Delta g = \left(\dfrac{\lambda c}{1+\gamma\lambda c}\right)(p_H + g_L)$, *then* $\beta^* \in \left[\dfrac{1}{1+\gamma\lambda c}, 1\right]$.

(iii) If $0 < \Delta p - \Delta g < \left(\dfrac{\lambda c}{1+\gamma\lambda c}\right)(p_H + g_L)$, *then* $\beta^* = 1$.

(iv) If $\Delta p - \Delta g = 0$, *then* $\beta^* = 1$.

(v) If $\Delta p - \Delta g < 0$ *and* $d \le \dfrac{1}{1+(1-\gamma)\lambda c}$, *then* $\beta^* = 1$.

(vi) If $\Delta p - \Delta g < 0$, $\left(\dfrac{1 - \gamma\lambda c}{\lambda c}\right)(\Delta g - \Delta p) < p_H + g_L$, and $d > \dfrac{1}{1 + (1 - \gamma)\lambda c}$, then $\beta^* = 1$.

(vii) If $\Delta p - \Delta g < 0$, $\left(\dfrac{1 - \gamma\lambda c}{\lambda c}\right)(\Delta g - \Delta p) = p_H + g_L$, and $d > \dfrac{1}{1 + (1 - \gamma)\lambda c}$, then

$$\beta^* \in \left[\frac{1}{d\left[1 + (1 - \gamma)\lambda c\right]}, 1\right].$$

(viii) If $\Delta p - \Delta g < 0$, $\left(\dfrac{1 - \gamma\lambda c}{\lambda c}\right)(\Delta g - \Delta p) > p_H + g_L$, and $d > \dfrac{1}{1 + (1 - \gamma)\lambda c}$, then

$$\beta^* = \frac{1}{d\left[1 + (1 - \gamma)\lambda c\right]}.$$

In cases (i)-(iii), $\alpha_P^* = \min\left\{(1 - \gamma) + \dfrac{1 - \beta}{\lambda\beta c}, 1\right\}$.

In case (iv), $\alpha_P^* \in \left[\max\left\{0, (1 - \gamma) - \dfrac{1 - d}{\lambda d c}\right\}, 1 - \gamma\right]$.

In cases (v)-(viii), $\alpha_P^* = \max\left\{0, (1 - \gamma) - \dfrac{1 - \beta d}{(\lambda\beta d)c}\right\}$.

In all eight cases, $\alpha_G^* = 1 - \alpha_P$.

Other Transaction (OT) Authority Reference

This appendix contains excerpts from the following four public laws pertaining to OT Authority:

- Public Law 101-189, National Defense Authorization Act for Fiscal Years 1990 and 1991, Section 251, Allied Cooperative Research and Development, November 29, 1989.
- Public Law 103-160, National Defense Authorization Act for Fiscal Year 1994, Section 845, Other Transactions (OTs) for Prototype Projects, November 30, 1993.
- Public Law 104-201, National Defense Authorization Act for Fiscal Year 1997, Section 804, Modification of Authority to Carry Out Certain Prototype Projects, September 23, 1996.
- Public Law 106-398, Floyd D. Spence National Defense Authorization Act for Fiscal Year 2001, Section 803, Clarification and Extension of Authority to Carry Out Certain Prototype Projects, October 30, 2000.

Public Law 101-189 §251

Bill Text
101st Congress (1989-1990)
H.R.2461.PP

THIS SEARCH	THIS DOCUMENT	GO TO
Next Hit	Forward	New Bills Search
Prev Hit	Back	HomePage
Hit List	Best Sections	Help
	Contents Display	

H.R.2461

National Defense Authorization Act for Fiscal Years 1990 and 1991 (Public Print - PP)

SEC. 251. ALLIED COOPERATIVE RESEARCH AND DEVELOPMENT

(a) DESIGNATION OF SUBCHAPTERS- Chapter 138 of title 10, United States Code, is amended--

> *(1) by striking out the chapter heading and inserting in lieu thereof the following:*

`CHAPTER 138--COOPERATIVE AGREEMENTS WITH NATO ALLIES AND OTHER COUNTRIES

`Subchapter

- `I.

-Acquisition and Cross-Servicing Agreements

-2341

- `II.

-Other Cooperative Agreements

-2350a

`SUBCHAPTER I--ACQUISITION AND CROSS-SERVICING AGREEMENTS'; and

> *(2) by adding at the end the following:*

`SUBCHAPTER II--OTHER COOPERATIVE AGREEMENTS

`Sec.

`2350a. Allied cooperative research and development.

`**Sec. 2350a. Allied cooperative research and development**

`*(a) AUTHORITY TO ENGAGE IN COOPERATIVE R & D PROJECTS- The Secretary of Defense may enter into a memorandum of understanding (or other formal agreement) with one or more major allies of the United States for the purpose of conducting cooperative research and development projects on defense equipment and munitions.*

`*(b) RESTRICTIONS- (1) A memorandum of understanding (or other formal agreement) to conduct a cooperative research and development project under this section may not be entered into unless the Secretary of Defense determines that the proposed project will improve through the application of emerging technology the conventional defense capabilities of the North Atlantic Treaty Organization (NATO) or the common conventional defense capabilities of the United States and its major non-NATO allies.*

`*(2) Each cooperative project entered into under this section shall require sharing of the costs of research and development between the participants on an equitable basis.*

`*(3) The Secretary may not delegate the authority to make a determination under paragraph (1) except to the Deputy Secretary of Defense or the Under Secretary of Defense for Acquisition.*

`*(c) RESTRICTIONS ON PROCUREMENT OF EQUIPMENT AND SERVICES- (1) In order to assure substantial participation on the part of the major allies of the United States in approved cooperative research and development projects, funds made available for such projects may not be used to procure equipment or services from any foreign government, foreign research organization, or other foreign entity.*

`*(2) A major ally of the United States may not use any military or economic assistance grant, loan, or other funds provided by the United States for the purpose of making that ally's contribution to a cooperative research and development program entered into with the United States under this section.*

`*(d) COOPERATIVE OPPORTUNITIES DOCUMENT- (1)(A) In order to ensure that opportunities to conduct cooperative research and development projects are considered during the early decision points in the Department of Defense's formal development review process in connection with any planned project of the Department of Defense, the Under Secretary of Defense for Acquisition shall prepare a formal arms cooperation opportunities document for review by the Defense Acquisition Board at its formal meetings.*

`*(B) The Under Secretary shall also prepare an arms cooperation opportunities document for review of each new project for which a Mission Need Statement is prepared.*

`*(2) The formal arms cooperation opportunities document referred to in paragraph (1) shall include the following:*

`*(A) A statement indicating whether or not a project similar to the one under consideration by the Department of Defense is in development or production by one or more of the major allies of the United States.*

`(B) If a project similar to the one under consideration by the Department of Defense is in development or production by one or more major allies of the United States, an assessment by the Under Secretary of Defense for Acquisition as to whether that project could satisfy, or could be modified in scope so as to satisfy, the military requirements of the project of the United States under consideration by the Department of Defense.

`(C) An assessment of the advantages and disadvantages with regard to program timing, developmental and life cycle costs, technology sharing, and Rationalization, Standardization, and Interoperability (RSI) of seeking to structure a cooperative development program with one or more major allies of the United States.

`(D) The recommendation of the Under Secretary of Defense for Acquisition as to whether the Department of Defense should explore the feasibility and desirability of a cooperative development program with one or more major allies of the United States.

`(e) REPORTS TO CONGRESS- (1) Not later than March 1 of each year, the Under Secretary of Defense for Acquisition shall submit to the Committees on Armed Services and Appropriations of the Senate and House of Representatives a report--

`(A) describing the status, funding, and schedule of existing cooperative research and development projects carried out under this section for which memoranda of understanding (or other formal agreements) have been entered into; and

`(B) describing the purpose, funding, and schedule of any new cooperative research and development projects proposed to be carried out under this section (including those projects for which memoranda of understanding (or other formal agreements) have not yet been entered into) for which funds have been included in the budget submitted to Congress pursuant to section 1105 of title 31 for the fiscal year following the fiscal year in which the report is submitted.

`(2) The Secretary of Defense and the Secretary of State, whenever they consider such action to be warranted, shall jointly submit to the Committees on Armed Services and Foreign Relations of the Senate and to the Committees on Armed Services and Foreign Affairs of the House of Representatives a report--

`(A) enumerating those countries to be added to or deleted from the existing designation of countries designated as major non-NATO allies for purposes of this section; and

`(B) specifying the criteria used in determining the eligibility of a country to be designated as a major non-NATO ally for purposes of this section.

`(f) SIDE-BY-SIDE TESTING- (1) It is the sense of Congress--

`(A) that the Department of Defense should perform more side-by-side testing of conventional defense equipment manufactured by the United States and other member nations of NATO; and

`(B) that such testing should be conducted at the late stage of the development

process when there is usually only a single United States prime contractor.

`(2) The Deputy Director of Defense Research and Engineering (Test and Evaluation) may acquire items of the type specified in paragraph (3) manufactured by other member nations of NATO for side-by-side comparison testing with comparable items of United States manufacture.*

`(3) Items that may be acquired by the Deputy Director under paragraph (2) include the following:*

  `(A) Submunitions and dispensers.*

  `(B) Anti-tank and anti-armor guided missiles.*

  `(C) Mines, for both land and naval warfare.*

  `(D) Runway-cratering devices.*

  `(E) Torpedoes.*

  `(F) Mortar systems.*

  `(G) Light armored vehicles and major subsystems thereof.*

  `(H) Utility vehicles.*

  `(I) High-velocity anti-tank guns.*

  `(J) Short-Range Air Defense Systems (SHORADS).*

  `(K) Mobile air defense systems and components.*

`(4) The Deputy Director shall notify the committees on Armed Services and on Appropriations of the Senate and House of Representatives of his intent to obligate funds made available to carry out this subsection not less than 30 days before such funds are obligated.*

`(5) Not later than February 1 of each year, the Deputy Director shall submit to the Committees on Armed Services and on Appropriations of the Senate and House of Representatives a report--*

  `(A) on the systems, subsystems, and munitions produced by other member nations of NATO that were evaluated during the previous fiscal year by the Deputy Director; and*

  `(B) on the obligation of any funds under this subsection during the previous fiscal year.*

`(g) SECRETARY TO ENCOURAGE SIMILAR PROGRAMS- The Secretary of Defense shall encourage major allies of the United States to establish programs similar to the one provided for in this section.*

`(h) DEFINITIONS- In this section:*

`(1) The term `cooperative research and development project' means a project involving joint participation by the United States and one or more major allies of the United States under a memorandum of understanding (or other formal agreement) to carry out a joint research and development program--

`(A) to develop new conventional defense equipment and munitions; or

`(B) to modify existing military equipment to meet United States military requirements.

`(2) The term `major ally of the United States' means a member nation of the North Atlantic Treaty Organization (other than the United States) or a major non-NATO ally.

`(3) The term `major non-NATO ally' means a country (other than a member nation of the North Atlantic Treaty Organization) designated as a major non-NATO ally for purposes of this section by the Secretary of Defense with the concurrence of the Secretary of State.'.

(b) REPEALS- Section 1103 of the Department of Defense Authorization Act, 1986 (10 U.S.C. 2407 note), and section 1105 of the Department of Defense Authorization Act, 1987 (22 U.S.C. 2767a), are repealed.

Public Law 103-160 §845

SEC. 845. AUTHORITY OF THE ADVANCED RESEARCH PROJECTS AGENCY TO CARRY OUT CERTAIN PROTOTYPE PROJECTS.

(a) AUTHORITY.—The Director of the Advanced Research Projects Agency may, under the authority of section 2371 of title 10, United States Code, carry out prototype projects that are directly

relevant to weapons or weapon systems proposed to be acquired or developed by the Department of Defense.

(b) EXERCISE OF AUTHORITY.—(1) Subsections (c)(2) and (c)(3) of such section 2371, as redesignated by section 827(b)(1)(B), shall not apply to projects carried out under subsection (a).

(2) The Director shall, to the maximum extent practicable, use competitive procedures when entering into agreements to carry out projects under subsection (a).

(c) PERIOD OF AUTHORITY.—The authority of the Director to carry out projects under subsection (a) shall terminate 3 years after the date of the enactment of this Act.

Public Law 104-201 §804

SEC. 804. MODIFICATION OF AUTHORITY TO CARRY OUT CERTAIN PROTOTYPE PROJECTS.

(a) AUTHORIZED OFFICIALS.—(1) Subsection (a) of section 845 of the National Defense Authorization Act for Fiscal Year 1994 (Public Law 103–160; 107 Stat. 1721; 10 U.S.C. 2371 note) is amended by inserting ", the Secretary of a military department, or any other official designated by the Secretary of Defense" after "Agency".

(2) Subsection (b)(2) of such section is amended to read as follows:

"(2) To the maximum extent practicable, competitive procedures shall be used when entering into agreements to carry out projects under subsection (a).".

(b) EXTENSION OF AUTHORITY.—Subsection (c) of such section is amended by striking out "terminate" and all that follows and inserting in lieu thereof "terminate at the end of September 30, 1999.".

(c) CONFORMING AND TECHNICAL AMENDMENTS.—Section 845 of such Act is further amended—

    (1) in subsection (b)—

        (A) in paragraph (1), by striking out "(c)(2) and (c)(3) of such section 2371, as redesignated by section 827(b)(1)(B)," and inserting in lieu thereof "(e)(2) and (e)(3) of such section 2371"; and

        (B) in paragraph (2), by inserting after "Director" the following: ", Secretary, or other official"; and

    (2) in subsection (c), by striking out "of the Director".

Public Law 106-398 §803

SEC. 803. CLARIFICATION AND EXTENSION OF AUTHORITY TO CARRY OUT CERTAIN PROTOTYPE PROJECTS.

(a) AMENDMENTS TO AUTHORITY.—Section 845 of the National Defense Authorization Act for Fiscal Year 1994 (Public Law 103–160; 10 U.S.C. 2371 note) is amended—

    (1) by redesignating subsection (d) as subsection (f); and

    (2) by inserting after subsection (c) the following new subsections:

"(d) APPROPRIATE USE OF AUTHORITY.—(1) The Secretary of Defense shall ensure that no official of an agency enters into a transaction (other than a contract, grant, or cooperative agreement) for a prototype project under the authority of this section unless—

    "(A) there is at least one nontraditional defense contractor participating to a significant extent in the prototype project; or

---

"(B) no nontraditional defense contractor is participating to a significant extent in the prototype project, but at least one of the following circumstances exists:

"(i) At least one third of the total cost of the prototype project is to be paid out of funds provided by parties to the transaction other than the Federal Government.

"(ii) The senior procurement executive for the agency (as designated for the purposes of section 16(3) of the Office of Federal Procurement Policy Act (41 U.S.C. 414(3)) determines in writing that exceptional circumstances justify the use of a transaction that provides for innovative business arrangements or structures that would not be feasible or appropriate under a contract.

"(2)(A) Except as provided in subparagraph (B), the amounts counted for the purposes of this subsection as being provided, or to be provided, by a party to a transaction with respect to a prototype project that is entered into under this section other than the Federal Government do not include costs that were incurred before the date on which the transaction becomes effective.

"(B) Costs that were incurred for a prototype project by a party after the beginning of negotiations resulting in a transaction (other than a contract, grant, or cooperative agreement) with respect to the project before the date on which the transaction becomes effective may be counted for purposes of this subsection as being provided, or to be provided, by the party to the transaction if and to the extent that the official responsible for entering into the transaction determines in writing that—

"(i) the party incurred the costs in anticipation of entering into the transaction; and

"(ii) it was appropriate for the party to incur the costs before the transaction became effective in order to ensure the successful implementation of the transaction.

"(e) NONTRADITIONAL DEFENSE CONTRACTOR DEFINED.—In this section, the term 'nontraditional defense contractor' means an entity that has not, for a period of at least one year prior to the date that a transaction (other than a contract, grant, or cooperative agreement) for a prototype project under the authority of this section is entered into, entered into or performed with respect to—

"(1) any contract that is subject to full coverage under the cost accounting standards prescribed pursuant to section 26 of the Office of Federal Procurement Policy Act (41 U.S.C. 422) and the regulations implementing such section; or

"(2) any other contract in excess of $500,000 to carry out prototype projects or to perform basic, applied, or advanced research projects for a Federal agency, that is subject to the Federal Acquisition Regulation.".

(b) EXTENSION OF AUTHORITY.—Subsection (f) of such section, as redesignated by subsection (a)(1), is amended by striking "September 30, 2001" and inserting "September 30, 2004".

References

"2012 Global R&D Funding Forecast," *R&D Magazine*, December 2011.

AlterG, *Learn More About Us*, web page, undated. As of December 6, 2012:
http://www.alterg.com/alter-g

Benjamin, T., *Venture Capital Concept Analysis, Final Report*, Homeland Security Institute, December 2005.

Britt, R., *Universities Report Highest-Ever R&D Spending of $65 Billion in FY 2011*, National Science Foundation, InfoBrief NSF 13-305, November 2012.

Brown, C., P. Winka, and H. Lee, *Government Venture Capital: Centralized or Decentralized Execution*, Naval Postgraduate School, NPS-AM-07-052, January 17, 2008.

Business Executives for National Security, *Accelerating the Acquisition and Implementation of New Technologies for Intelligence: The Report of the Independent Panel on the Central Intelligence Agency In-Q-Tel Venture*, June 2001.

Caterinicchia, D., "NIMA Tries Venture Capital Route," *Federal Computer Weekly*, March 24, 2002.

Defense Advanced Research Projects Agency, *DARPA Network Challenge*, web page, undated. As of October 13, 2012:
http://archive.darpa.mil/networkchallenge/

Defense Venture Catalyst Initiative, *Department of Defense Launches the Defense Venture Catalyst Initiative to Speed Discovery of Emerging Commercial Technologies*, undated. As of December 17, 2012:
http://web.archive.org/web/20061018215004/http://devenci.dtic.mil/pdf/DeVenCIAnnouncement.pdf

Defense Venture Catalyst Initiative, *Frequently Asked Questions*, undated. As of August 23, 2012:
http://web.archive.org/web/20110814135817/http://devenci.dtic.mil/faq.htm

———, *Welcome to Defense Venture Catalyst Initiative*, undated. As of August 23, 2012:
http://web.archive.org/web/20120118055250/http://devenci.dtic.mil/

Department of Agriculture, Office of the Inspector General, *Assessment of the Alternative Agricultural Research and Commercial Corporation–Management Lacking Over High Risk Investments*, Audit Report No. 37099-1-FM, November 1999.

Department of Defense, *Annual Report on Cooperative Agreements and Other Transactions Entered into During FY2003 Under 10 USC 2371*, undated a.

———, *Annual Report on Cooperative Agreements and Other Transactions Entered into During FY2004 Under 10 USC 2371*, undated b.

Drezner, J., G. Smith, and I. Lachow, *Assessing the Use of "Other Transactions" Authority for Prototype Projects*, Santa Monica, Calif.: RAND Corporation, DB-375-OSD, 2002. As of August 6, 2013:
http://www.rand.org/pubs/documented_briefings/DB375.html

Dunn, R., *Injecting New Ideas and New Approaches in Defense Systems: Are "Other Transactions" an Answer?* Naval Postgraduate School, Annual Acquisition Research Conference, May 2009.

Fike, G., "Measuring 'Other Transaction' Authority Performance Versus Traditional Contracting Performance: A Missing Link to Further Acquisition Reform," *The Army Lawyer*, July 2009.

The Foundation Center, *Change in Foundation Giving Adjusted for Inflation, 1975 to 2011*, FC Stats: The Foundation Center's Statistical Information Service, 2013.

Graham, D. R., J. P. Bell, and A. J. Coe, *Defense Venturing Process: A Model for Engaging Venture Capitalists and Innovative Emerging Companies*, Institute for Defense Analyses, D-2847, March 2003.

Halchin, L. E., *Other Transaction (OT) Authority*, Congressional Research Service, RL34760, January 27, 2010.

Held, B., and I. Chang, *Using Venture Capital to Improve Army Research and Development*, Santa Monica, Calif.: RAND Corporation, IP-199, 2000. As of August 6, 2013:
http://www.rand.org/pubs/issue_papers/IP199.html

Higgins, K. J., "'Mudge' Announces New DARPA Hacker Spaces Program," *DarkReading.com*, August 4, 2011. As of August 6, 2013:
http://www.darkreading.com/security/news/231300269/mudge-announces-new-darpa-hacker-spaces-program.html

Khalili, O., *15 Social Venture Capital Firms That You Should Know About*, Cause Capitalism, April 2, 2010. As of August 6, 2013:
http://causecapitalism.com/15-social-venture-capital-firms-that-you-should-know-about/

Laurent, A., "Raising the Ante: Venture Capitalists Are Helping Government Buy Its Way Back into the Emerging Technology Market," *Government Executive*, June 1, 2002. As of August 6, 2013:
http://www.govexec.com/magazine/2002/06/raising-the-ante/11584/

Mara, A. S., *Maximizing the Returns of Government Venture Capital Programs*, National Defense University, Defense Horizons, January 2011.

MassVentures, *History*, web page, undated. As of August 6, 2013:
http://mass-ventures.com/about_us/history

Mawsonia Limited 2010, *Global Corporate Venturing Annual Review 2012*, Issue 031, December 2012.

McBride, A., *Pentagon Turns to Silicon Valley for Leads*, Reuters US Online Report, Technology News, October 14, 2011. As of September 13, 2013:
http://www.reuters.com/article/2011/10/14/us-venture-pentagon-idUSTRE79D60J20111014

Molzahn, W., "The CIA's In-Q-Tel Model – Its Applicability," *Acquisition Review Quarterly*, Winter 2003.

Nalebuff, B. J., and J. E. Stiglitz, "Prizes and Incentives: Towards a General Theory of Compensation and Competition," *The Bell Journal of Economics*, Vol. 14, No. 1, Spring 1983, pp. 21–43.

National Public Radio, *In-Q-Tel: The CIA's Tax-Funded Player in Silicon Valley*, All Tech Considered transcript, July 16, 2012. As of August 6, 2013:
http://www.npr.org/templates/transcript/transcript.php?storyId=156839153

National Science Board, *Research and Development: Essential Foundation for U.S. Competitiveness in a Global Economy*, Arlington, VA, 08-03, January 2008.

National Science Foundation, *National Patterns of R&D Resources: 2010–11 Data Update, Detailed Statistical Tables*, NSF 13-318, April 2013.

Office of the Assistant Secretary of the Army for Financial Management and Comptroller, *Source of Funds for Army Use (Other than Typical Army Appropriations)*, Resource Analysis and Business Practices, SAFM-RB, March 2005.

Office of the Under Secretary of Defense for Acquisition, Technology, and Logistics, *Fulfillment of Urgent Operational Needs*, Report of the Defense Science Board Task Force, July 2009.

OnPoint Technologies, *Portfolio*, web page, undated. As of May 26, 2013:
http://www.onpoint.us/portfolio/index.shtml#

Pratt, A., *Innovation for the U.S. Intelligence Community*, Summer Intern Blog, Stanford Graduate School of Business, July 14, 2011. As of August 6, 2013:
http://csi.gsb.stanford.edu/innovation-us-intelligence-community

Prequin, Ltd., *2012 Prequin Global Private Equity Report*, 2012.

Public Law 101-189, National Defense Authorization Act for Fiscal Years 1990 and 1991, Section 251, Allied Cooperative Research and Development, November 29, 1989.

Public Law 103-160, National Defense Authorization Act for Fiscal Year 1994, Section 845, Other Transactions (OTs) for Prototype Projects, November 30, 1993.

Public Law 104-201, National Defense Authorization Act for Fiscal Year 1997, Section 804, Modification of Authority to Carry Out Certain Prototype Projects, September 23, 1996.

Public Law 106-398, Floyd D. Spence National Defense Authorization Act for Fiscal Year 2001, Section 803, Clarification and Extension of Authority to Carry Out Certain Prototype Projects, October 30, 2000.

Reardon, M., and D. Scott, *Rosettex NTA Project Portfolio, Final Edition*, Sarnoff Corporation, National Technology Alliance, TR-001-072709-554, September 2009.

Red Planet Capital, *Investment Sectors*, web page, undated. As of November 19, 2012:
http://web.archive.org/web/20061106032439/http://www.redplanetcapital.com/index.php?pid=23

Rosettex Technology and Ventures Group, *Rosettex and NTA*, web page, 2004. As of October 29, 2012:
http://web.archive.org/web/20080828041219/http://www.rosettex.com/nta/index.asp

Scotchmer, S., *Innovation and Incentives*, Cambridge, Mass.: MIT Press, 2004.

Seebold, M., *Saffron Technology, Part of the Rosettex Technology and Ventures Group, Wins Major Award from US Government's National Imagery and Mapping Agency (NIMA)*, Saffron Technology press release, May 2, 2002.

Shell Foundation, *Establishing Entities*, web page, 2008. As of June 15, 2012:
http://shellfoundation.org/pages/core_lines.php?p=our_approach_content&page=establish

Sidebottom, D., *Innovative Contracting Methods*, briefing, April 19, 2010. As of November 7, 2012:
http://www.defense.gov/Blog_files/Blog_assets/Presentations/Pres_OtherTransactionsOpenInnovationsOGC OTBriefing_19Apr10.ppt

Smith, D., *USDA Investments at Risk Due to Corporation's Mismanagement*, U.S. Department of Agriculture press release, December 2, 1999.

Snider, L. B., *The Agency and the Hill: CIA's Relationship with Congress, 1946–2004*, Washington, D.C.: The Center for the Study of Intelligence, Central Intelligence Agency, 2008. As of September 16, 2013:
https://www.cia.gov/library/center-for-the-study-of-intelligence/csi-publications/books-and-monographs/ agency-and-the-hill/

Sobel, D., *Longitude: The True Story of a Lone Genius Who Solved the Greatest Scientific Problem of His Time*, New York: Walker and Company, 1995.

"Social Impact Bonds: Commerce and Conscience," *The Economist*, February 23, 2013. As of September 13, 2013:
http://www.economist.com/news/finance-and-economics/21572231-new-way-financing-public-services-gains-momentum-commerce-and-conscience

Steitz, D., and P. Banks, *NASA Forms Partnership with Red Planet Capital, Inc.*, News Release 06-317, September 20, 2006. As of November 14, 2012:
http://solarsystem.nasa.gov/news/display.cfm?News_ID=16360

Taulli, T., "The Lowdown on Strategic Investments," *Bloomberg Businessweek*, July 2, 2008.

UNU-MERIT, *Study on the Economic Impact of Open Source Software on Innovation and the Competitiveness of the Information and Communication Technologies (ICT) Sector in the EU*, Final Report, November 20, 2006.

Yannuzzi, R., "In-Q-Tel: A New Partnership Between the CIA and the Private Sector," *Defense Intelligence Journal*, Winter 2000.